Robert Henry Scott

Elementary Meteorology

Second Edition

Robert Henry Scott

Elementary Meteorology

Second Edition

ISBN/EAN: 9783337279837

Printed in Europe, USA, Canada, Australia, Japan

Cover: Foto ©berggeist007 / pixelio.de

More available books at **www.hansebooks.com**

ELEMENTARY
‘METEOROLOGY·

BY

ROBERT H. SCOTT, M.A., F.R.S.

SECRETARY TO THE METEOROLOGICAL COUNCIL, AND AUTHOR OF
'WEATHER CHARTS AND STORM WARNINGS'

WITH NUMEROUS ILLUSTRATIONS

SECOND EDITION

LONDON
KEGAN PAUL, TRENCH, & CO., 1 PATERNOSTER SQUARE
1883

PREFACE.

APPLICATION has been so frequently made to me to recommend a simple text-book of Meteorology, that I have been induced to compile the following pages in the endeavour to supply the want.

This book lays but little claim to originality, and in the first place my sincerest acknowledgments are due to my friend Dr. Julius Hann, of Vienna, whose recently published 'Erdkunde' has been freely used in the preparation of Part II. Professor Mohn's 'Grundzüge der Meteorologie' and Buchan's 'Handy-book of Meteorology' have also been frequently laid under contribution.

If, in any case, full credit has not been given to the first discoverer of any principle, I must only beg him to accept my apologies for the omission.

In conclusion, my warm thanks must be expressed to several kind friends who have taken the trouble of reading over the proof-sheets.

ROBERT H. SCOTT.

METEOROLOGICAL OFFICE:
December, 1882.

CONTENTS.

PART I.

CHAPTER   PAGE

I. INTRODUCTORY 1

    Meteorology, as a science, contrasted with astronomy, 1.—Subdivision of the subject, 3. Climatology, 3. Weather study, 3. Cosmical meteorology, 4.—Plan of the work, 5.

II. THE EARTH AND THE ATMOSPHERE 6

    The solar system, 6. The earth's motion, 7.—Distribution of land and water, 11.—The atmosphere, 12; its constitution, 13.

III. TEMPERATURE 17

    The thermometer, 17; its construction, 17. Scales, 19. Fixed points, 21.—Various thermometers—ordinary, 23; self-registering, 25; maximum, 26; minimum, 29.—Thermographs, 32.—Thermometric exposure, 33. Screens, 34.— Mean temperature, 37. Hours of observation, 39.—Diurnal and annual range, 40. Causes which produce them, 45. Chrono-isothermal diagram for Greenwich, 48.

IV. RADIATION 50

    Sources of heat, 50. Solar radiation, 51. Actinometers, 52. Black bulb thermometers, 53. Absorption of heat by atmosphere, 56. Violle's results, 57.—The sunshine recorder, 58.
    Terrestrial radiation, 59. Grass temperatures, 60. Formation of ice by radiation in India, 61.

CONTENTS.

CHAPTER	PAGE
V. PRESSURE	63

Invention of the barometer, 63; its construction, 65. The vernier, 67.—Reading the barometer, 68. Its management, 71. Different forms of barometers—Fortin's, 72; Kew pattern, 73; marine, 74; siphon, 75.—Barographs, 77.—Aneroids, 79.—Sympiezometers, 81.—Defects of the barometer, 82. The pipette, 83.—Barometrical corrections, 84. Reduction to sea-level, 85.—Discussion of the observations, 88.—Range curves, diurnal and annual, 89.

VI. THE MOISTURE OF THE ATMOSPHERE 94

Atmometry, 95. The difference between gases and vapours, 95. Evaporation, 96; its amount in different latitudes, 101. Hygrometry, 103. Direct hygrometers, 103. Daniell's, 103. Regnault's, 104. Dines's, 104.—Indirect hygrometers, 105. Organic—Saussure's, 105. Inorganic—the dry and wet bulb thermometers, 107.—Formula for deducing results, 108. The Greenwich factors, 109.
Distribution of vapour in vertical height, 111; its action on radiant heat, 113.

VII. DEW, FOG, MIST, AND CLOUD 114

Wells' theory of dew, 114. Hoar frost, 115. Glazed frost, 115. Prediction of hoar frost, 117. Estimates of amount of dew, 118.
Condensation, 119. Fogs, 120. Causes which produce them, 120.
Clouds, 123. Howard's classification, 124. Their motion, 129 —amount, 129.

VIII. RAIN, SNOW, AND HAIL 131

Rain, 131. Rain gauges, 131. Their exposure, 133. Influence of height on their indications, 133. The British rainfall system, 135. Storage of rain, 135.—Floods, 136. Excessive falls of rain, 137.—Causes which produce rain, 137.
Snow, 141; its collection, 141; measurement, 141. Forms of snow crystals, 142.
Hail, 143. Forms of hail-stones, 143. Theories of their formation, 145.
Diurnal range of rainfall, 146.

CONTENTS. ix

CHAPTER PAGE
IX. WIND 148
Direction of the wind, 148. Anemometers, 150.—Pressure, 150.—Velocity, 151. Conversion of velocity into pressure, 154. Extreme pressures recorded, 156. Beaufort's scale of wind force. 158.—Discussion of wind observations, 161. Lambert's formula, 161. Diurnal and annual variations of velocity, 163. Wind components, 165.

X. ELECTRICAL PHENOMENA 167
Franklin's experiment, 167. Electrometers, 168. Thomson's electrograph, 169. Results of observations of atmospheric electricity at Brussels, 171. Phenomena during thunderstorms, 172.
Lightning 173; zigzag, 173; sheet, 175; globular, 175. St. Elmo's fire, 178.—Thunder, 178.—Comparison of lightning with the electric spark, 180. Action of lightning, 181. The return shock, 181. Lightning conductors, 183. Rules for their erection, 186. The Lightning Rod Conference, 186.
Thunderstorms, 188. Classification, 189. Diurnal and annual period, 190.
The Aurora, 191. Estimates of its height, 193; connection with weather, 195; with terrestrial magnetism, 196.
Ozone, 196; difficulty of observing it, 197.

XI. OPTICAL PHENOMENA 199
The rainbow, 199; primary. 200; secondary, 200; supernumerary, 200; extraordinary, 201.—Fog bows, 201. Coronæ, 202.—Glories, 203.—Halos, 203.—Parhelia, 203. Cloud colouring, 205.—Mirage, 206.—Looming, 207.

PART II.

XII. THE DISTRIBUTION OF TEMPERATURE 211
The action of the sun's rays, 211; influences diurnal range, 214. Annual range, 218. Influence of land and water respectively, 219. Action of ocean currents, 221.
Isothermal charts, 222.—Isabnormal charts, 232.—Extremes of temperature, 235.—Temperature at high elevations, 235.

a

APPENDICES.

		PAGE
1. Comparison of thermometer scales	. . . . . .	389
2. Mode of construction of the chrono-isothermal diagram.	. .	390
3. Correction of barometers for temperature	. . . .	391
4. ,, ,, altitude	. . . . .	391
5. Note on the relation between sunspots and weather	. .	392

INDEX 395

LIST OF PLATES.

(To be placed at end.)

I. Isothermal lines, showing the mean temperature of the globe for the year.

II. Isothermal lines, showing the mean temperature of the globe for January.

III. Isothermal lines, showing the mean temperature of the globe for July.

IV. Lines, showing the isabnormal temperatures of the globe in January.

V. Lines, showing the isabnormal temperatures of the globe in July.

VI. Map, showing the mean atmospheric pressure and prevailing winds of the globe for January.

VII. Map, showing the mean atmospheric pressure and prevailing winds of the globe for July.

VIII. Current chart of the globe for the year.

IX. Isothermal lines, showing the mean sea surface temperature for February.

X. Isothermal lines, showing the mean sea surface temperature for August.

XI. Lines, showing the regions of equal annual range of temperature.

Errata

Page 26, line 18, *dele* 'the two ends of the mercury will indicate different temperatures, and'

„ 40, Fig. 7, the point R should coincide with the vertical line for 9 A.M., and the line R S be drawn accordingly

„ 69, Fig. 11A, the letter *y* is misplaced, and should stand two divisions lower on the scale

„ 102, line 9, *for* Cole *read* Collin

„ 107, „ 9, *for* Sir John Leslie *read* Hutton

„ 107, *dele* note, and substitute 'See *Trans. Royal Soc. Edinburgh*, Vol. V. Part 3, page 67, Playfair, Life of Hutton'

„ 115, number of page to be corrected from 155

„ 120, line 15, *for* Dr. Aitken *read* Mr. J. Aitken

„ 132, line 8, *for* devised by the Meteorological Office in 1868 *read* apparently first proposed by Quetelet in 1852[1]

„ 132, *insert note*:—[1] *Sur le Climat de la Belgique*, cinquième partie, p. 2. Brussels, 1852

„ 137, line 17, *for* St. Paul's Churchyard *read* Camberwell

„ 139, in note, *for* Hochstette *read* Hochstetter.

„ 162, Fig. 29, *insert* A B at the ends of the resultant line

„ 248, last line, *for* Moher *read* Mohn.

„ 255, *insert note*:—These plates are chiefly based on the maps of the distribution of pressure which appeared in *Aids to the Study and Forecast of Weather*, by the Rev. W. Clement Ley, published by the Meteorological Council, 1879.

„ 289, line 2, *for* and cold *read* and comparatively cold (see p. 213).

„ 289, line 5, *after* the air *insert* already containing the heat imparted to it by vapour condensation, as described at p. 213.

„ 300, line 5, *dele* 'as Rennell's current'

„ 302, „ 11, *for* Rennell's *read* The

„ 309, „ 10 from bottom, *for* Tegethoff *read* Tegetthoff

INDEX

„ 397, second col., *for* Cole *read* Collin

„ 398, first col., *under* Current, *dele* 'Rennell's, 302'

„ 404, second col., *under* Rennell, *dele* 'his current, 302'

„ 406, „ „ *for* Tegethof *read* Tegetthoff

CONTENTS.

APPENDICES.

ELEMENTARY METEOROLOGY.

PART I.

CHAPTER I.

INTRODUCTORY.

METEOROLOGY is the science of the Atmosphere, of τὰ μετέωρα, the things above the earth, as Aristotle has it, and its interest to everyone scarcely needs to be dwelt upon, for as without air we cannot live, any knowledge which we can gain of its condition from time to time, and of the changes which are taking place in it, cannot fail to be of the very highest importance to our health and comfort.

Almost everyone imagines himself a born meteorologist, and from the earliest times men have been watching the weather and its changes and recording their experience thereof; in fact, the Book of Job, believed to be one of the earliest of the Biblical Canon, contains some sound meteorological knowledge, as true now as it was some three thousand years ago.

Nevertheless, though men have studied meteorology more or less systematically ever since the time of

B

Aristotle, who wrote the first treatise on the subject, little progress was made in it until the invention of the barometer and thermometer, about 200 years ago, and we must admit that even yet it has hardly made good its title to a place among the exact sciences.

The reason of this is easily explained. Firstly, we live at the bottom of the atmospheric ocean, of which the upper layers are practically inaccessible, and their condition almost unknown to us. Secondly, the observations we make of the physical state of the air are affected to such a degree by local accidents, such as the elevation, contour, and slope of the ground, its nearness to the sea, and even the character of the soil, that we meet with considerable variations of meteorological circumstances even within the limits of a single county.

In this respect meteorology affords a strong contrast to astronomy. The objects of observation and study in the latter science are at such a distance from the earth that it is practically of minor importance whether they be observed from Greenwich, Rome, or Washington. The phenomena themselves are identical, and, other things being equal, the difficulties of making the observation depend mainly on the meteorological conditions of the locality. In fact, under favourable meteorological circumstances, the range of phenomena observable by an astronomer is limited solely by the horizon of his station, and the power of his telescope. But in meteorology itself the case is widely different; the phenomena are not the same at two different points of observation. To take a single element, the temperature of the air in the streets of London differs appreciably from that experienced at the same time in the middle of the parks, and à *fortiori* from that observed entirely outside the city, as at Kew or Greenwich.

Hence we see the necessity of covering the country with a network of independent meteorological stations, as the observer at each place cannot do much more than record the phenomena exhibited by the portion of the atmosphere actually in contact with his instruments.

We may exemplify the difference between the two sciences by an illustration taken from biology; the astronomer may be compared to some of the more highly organised among the mollusca, such as the octopus, which is endowed with powers of locomotion and can seek his food at a distance from his home; while the meteorologist is like a mussel or an oyster, anchored to one spot, and obliged to make the best of such nutriment as may chance to be swept within his reach.

Meteorology may be considered from many different points of view. In the first instance, observations taken systematically at one place give eventually information as to the climate of that place, and when the results obtained for one such station are combined with those for other stations and compared with those taken in other countries, deductions may be drawn bearing on the relative fitness of different localities for the support of animal and vegetable life, &c., &c.

From this climatological point of view, the subject is immediately related to the science of physical geography, and in a greater or less degree to sanitary science also, and it is in this connection that it has been longest and most perseveringly studied.

Secondly, meteorology may be treated as the science of weather: that is of the changes which are from time to time taking place in the physical condition of the atmosphere and of the effects produced by such changes. These effects find their expression in the temperature of the air, the direction and motion of the wind, the amount of moisture contained in the atmosphere, and the balancing of the antagonistic forces of evaporation and condensation, on which depend what is termed, in common parlance, the fineness, or the contrary, of the weather.

This branch of the subject has attracted much attention of late years, owing to the development of telegraphy and the facilities which are thereby afforded for examining the conditions of weather existing simultaneously at different places. We cannot, however, claim for the study of weather that it has as yet made much practical progress in enabling us to foresee or forecast its course for more than a few hours, and we must admit that it has made next to no progress at all in gaining an insight into the agencies which are at work in producing the various phases of weather.

Thirdly, we have the highest object of meteorology, if we consider it as a department of cosmical physics: the investigation of the physical conditions of the atmosphere, and their relations to the forces of light, heat, electricity, and magnetism. We cannot doubt that this department of meteorology, if thoroughly followed up, will yield results of the highest interest and importance, besides throwing light on many common phenomena, which are as yet entirely unexplained.

Here, however, we are at once staggered by the difficulty to which allusion has already been made. We live at the bottom of an ocean, and have no means of testing its condition above the level at which we live, while the actual observations taken at the different stations are affected to a most perplexing extent by the peculiarities of local circumstances.

The aim of the present work is to set before the reader the conditions requisite for the successful prosecution of the science from what we have defined as the first point of view, and to show some of the results which have been derived therefrom, without, however, trenching too much on the domain of physical geography. The subject of weather will be merely dealt with in a passing way, as to a certain extent foreign to the consideration of climatology; while the third department, that of cosmical meteorology, will not be considered at all, as it would be out of place in a book professedly elementary.

The plan of the work is as follows—In Part I., after a short preliminary notice of the earth itself and its atmosphere, the various instruments and the modes of using them will be described, for the entire superstructure of reasoning in meteorology rests on the foundation of accurate observation; and unless this be secured by careful and long-continued attention to a few simple and obvious principles, the labour bestowed on the most complete mathematical discussion of the results will be thrown away.

In Part II. a brief account will be given of the geographical distribution of the different phenomena, which will serve as a general introduction to the science of physical geography as explained in other text-books.

CHAPTER II.

THE EARTH AND THE ATMOSPHERE.

BEFORE proceeding to treat of the atmosphere which envelopes the earth, it is well to state a few elementary facts as to the earth itself, in its relation to the sun, the great source whence, practically, all the heat which comes to us is derived; and as to the distribution of land and water on the surface of the globe, for this materially influences climate; all of which are important factors in the production of the meteorological phenomena which it is our object to discuss.

The form of the earth is that of an oblate spheroid, of which the equatorial radius is 3963·3 miles and the polar radius 3949·5 miles.[1]

In the solar system the earth is one of the six known as the old planets, and of these it comes third in order from the sun. These planets divide themselves into two groups. The four inner ones, Mercury, Venus, the Earth, and Mars, are characterised by high mean densities, comparatively slight ellipticities, slow rotation on their axes (that is, days about 24 hours long), and the rarity of satellites.

The two outer planets, Jupiter and Saturn, have low mean densities, great ellipticities, rapid rotation on

[1] Beckett, *Astronomy without Mathematics*, p. 6.

their axes (that is, days about 10 hours long), and abundance of satellites, Jupiter possessing four, and Saturn no less than eight, besides his rings.

The limits of the outer group have been extended within the last hundred years by the discovery of two others, Uranus and Neptune, and these agree in their characters with Saturn and Jupiter.

The space between the orbits of Mars and Jupiter is occupied by the paths of numerous smaller planets or asteroids, of which more than a hundred have been discovered within the present century.

It is, however, exclusively with the relations between the sun and the earth that we have to deal.

Among the principles of planetary motion discovered by Kepler, the famous German astronomer, and known as Kepler's Laws, the first is that the planets move in ellipses, each ellipse having the sun in one of its foci.

In the case of the earth the ellipse does not differ much from a circle. The mean distance of the earth from the sun being 92,000,000 miles, its distance on Jan. 1 is 90,436,000 miles, and on July 1, 93,564,000 miles.[1] We see, therefore, that the earth is nearer to the sun in the northern winter than in its summer.

The length of the earth's orbit is 578,052,560 miles, and as that is passed over in a year, or in 365 days, 5 hours, 48 minutes, 47·6 seconds, the hourly rate of motion through space is 65,941 miles. No perceptible effect of this motion on meteorological conditions is observable, as the atmosphere is carried round with the earth in its progress.

The rotation of the earth on its axis is, however, a motion which exerts considerable influence on meteoro-

[1] Beckett, *Astronomy without Mathematics*, p. 80.

logical phenomena, by affecting the wind, as we shall see further on.

The earth rotates on its axis once in twenty-four hours, and, as its equatorial circumference is 24,900 miles, a simple calculation shows that a point at the equator must be carried round at the rate of 1,040 miles an hour. It is also evident that a point exactly at either pole will not be carried round at all, and that points intermediate between these two extremes will be carried round at a less rate the nearer they are to the pole. Thus in latitude 30° the motion will be 900 miles an hour, and in latitude 60° it will be only 520 miles an hour, or one-half what it is at the equator.

It is, however, the apparent motion of the sun which exerts the greatest influence in producing meteorological changes. The axis of the earth changes its direction in space very slowly, but in the course of a year the apparent path of the sun in the sky describes a circle, called the ecliptic; the plane of this circle cuts the plane of the equator, supposed to be extended to the heavens, at the angle of 23° 27′ 44″, or, in round numbers, $23\frac{1}{2}°$. These circles accordingly intersect each other in two points diametrically opposite to each other. The sun appears on the equator on March 21, when he is going northwards, and on September 22, when he is going southwards. On June 21 he is farthest from the equator on the northern side, when he attains his greatest northern declination of $23\frac{1}{2}°$. On December 22 he attains his greatest southern declination of $23\frac{1}{2}°$ S.

When the sun is on the equator he is above the horizon for twelve hours, and below it for an equal time. This is expressed in common language by saying that the day and night are equal, and these epochs are called the equinoxes. When the sun reaches his extreme northern or southern declination, and commences his return southwards or northwards, as the case may be, he appears to stand still in the sky, and these epochs are called the solstices. In the latitude of London the length of the longest day, at the summer solstice, is 16 hours 34 minutes, and of the shortest day, at the winter solstice, 7 hours 47 minutes.

The proportion between the longest and shortest days depends on the latitude, and this brings us to the subject of the zones of geography: a division first proposed by Parmenides, a friend and older contemporary of Socrates, who lived about 450 B.C.

Parmenides divided the earth into three zones—Torrid, Temperate, and Frigid, and taught that the temperate zone was the only habitable region.

The modern definition of zones has reference to the position of the sun in the ecliptic, in relation to the successive parallels of latitude.

As the sun travels from $23\frac{1}{2}°$ N. to $23\frac{1}{2}°$ S. it must stand vertical, at some time or other, over every portion of the globe which lies between these limits of latitude. This belt, 47° in width, is called the Torrid Zone, and the lines bounding it are called the Tropics.

Within the Torrid Zone there is always true day and night, and the length of the days is reasonably uniform throughout the year.

On either side of the Torrid Zone we have a Temperate Zone, 43° in width, extending from either Tropic to the corresponding Polar Circle, in $66\frac{1}{2}°$ latitude. These circles are called respectively Arctic and Antarctic.

In the Temperate Zones there is always true day and night, but with differences in their length depending on the latitude, and attaining their greatest extent at the respective solstices. As the sun is theoretically visible for 90° from the horizon it is evident that, when he is at the winter solstice, in $23\frac{1}{2}$° S. latitude, he will be just on the horizon of an observer in $66\frac{1}{2}$° N. latitude, while to an observer further north he would be quite invisible, and that observer would have no true day.

The two zones lying round the poles, and bounded by the Arctic and Antarctic Circles respectively, are called the Frigid Zones, and within them during part of the year there is no day, and at the converse season, no night. At the poles themselves, strictly speaking, the day and night should be each of six months' duration, if it were not for the effect of atmospheric refraction, for the sun should be entirely invisible when he is south or north of the equator respectively.

Before we leave the subject of the zones we must remark that the interval from the spring to the autumn equinox is 184 days, while that from the autumn to the spring equinox is only 181. In other words, the sun remains three days longer over the northern than over the southern hemisphere. This ought to produce a difference in the amount of heat received from the sun in the respective hemispheres, but the compensation for this inequality in the length of the summer is found in the fact already explained, that, owing to the fact of the earth's orbit being an ellipse, not a circle, the sun is actually nearer to the earth in the summer of the southern hemisphere than in that of the northern, so that within the same period more heat is received by the southern than by the northern hemisphere. With the existing value of the eccentricity of the earth's orbit, the amount of heat received in perihelion (the southern summer) is to that received in aphelion (the northern summer) as 1·034 : 0·967.

There are two most important facts which must be remembered when dealing with the influence of the sun's heat in raising the temperature of the earth.

The first is, that, as water possesses a much higher specific heat and lower radiating power than land, the presence or absence of the sun's rays produces much less effect on the surface temperature of the oceans than of the continents.

The second is that the distribution of land and water on the globe is very uneven, and exhibits a remarkable arrangement. The computed ratio of land to water is generally stated at 1 : 3, but Lyell, in his 'Principles of Geology,' shows that the proportion of 1 : 4 is nearer the mark.

To exhibit the apparent capriciousness of the arrangement of the land it need only be said that the globe might be divided into two hemispheres, of which one should be covered almost entirely with water, while the other should contain more land than water. England would be situated near the centre of this latter hemisphere. If we mark on a globe all portions of land which have land antipodal to them, it will be found that they form only one twenty-seventh part of the existing dry land.

As regards the portions of land lying outside the tropics the ratio is also very remarkable, for there is thirteen times as much land in the northern as in the southern hemisphere.

I shall have to return to these subjects at greater length when treating of the distribution of temperature on the globe in Chapters XII. and XVII., and must now proceed to give some notice of the nature and constitution of the atmosphere.

The *Atmosphere* is a gaseous body surrounding the earth, and so distributed with regard to it as to form a complete envelope, of which the outermost shell is supposed to be similar to, and concentric with, the figure of the earth itself. The atmosphere, being material, is of course subject to the action of gravity, and, being gaseous, is endowed with the qualities of elasticity and great sensitiveness to the effects of heat, &c. Its condition, therefore, at any place is determined by the balance between the various forces acting on it, but necessarily it is much denser close to the earth than above that level. As is well known, at such heights as about seven miles, believed to have been attained by Messrs. Coxwell and Glaisher in their memorable balloon ascent from Wolverhampton, September 5, 1862, the greatest elevation ever reached, the atmosphere is so rarefied that great difficulty is experienced in breathing.

We may therefore suppose that the height of about seven miles is the limit at which mammals can exist, although, perhaps, some birds can endure a somewhat greater elevation. Such a height as even ten miles is, of course, very small as compared with the diameter of the earth, and utterly insignificant in relation to the distance between the earth and even the moon, not to speak of other heavenly bodies. In fact, on a globe 24 inches in diameter such a height would correspond to a shell 0·04 inch in thickness. The atmosphere,

THE EARTH AND THE ATMOSPHERE. 13

however, extends far beyond the highest level at which respiration can be sustained, but at a height of about 40 miles it is so rarefied as to be no longer capable of refracting the sun's rays, and its superior limit may, for practical purposes, be assumed at about 200 miles from the earth's surface. These estimates have been framed by M. Liais from observations on the phenomena of twilight at Rio Janeiro.[1]

Atmospheric air is not a simple, or even a compound, gas, but is essentially a mixture of the two gases, oxygen and nitrogen, in nearly the following proportions:—

	By Volume	By Weight
Nitrogen	79·1	76·9
Oxygen	20·9	23·1
	100·0	100·0

The statement that the air is not a chemical compound is proved by the following, among many other considerations:—Firstly, the proportions just cited bear no relation to the equivalent numbers of the respective gases, as would be the case if they were chemically combined; secondly, there are no signs of chemical action manifested when the gases are brought together, for, while pure nitrogen will not support combustion or respiration, as soon as about one-fourth of its bulk of oxygen has been added to a quantity of that gas, and the whole well shaken, the resulting mixture will support both processes as perfectly as the surrounding air of the atmosphere, while no indication of chemical action, such as the development of heat, has been visible.

It is generally asserted that the air is identical in

[1] *Comptes Rendus*, tome XLVIII., p. 109.

composition, whether it be taken from the surface of the earth or from the tops of the highest mountains, and for most practical purposes this may be assumed to be true; but, when we come to look into the matter more closely, we find that the assertion is only approximately correct, and that variations in the constitution of the air exist, which affect our comfort and our health.

Dr. Angus Smith[1] gives the following variations in the percentage amount of oxygen in specimens of air taken in different localities:—

An open heath in Scotland	20·9990
Open places in London in summer	20·9800
Court of Queen's Bench, February 2, 1866	20·6500
The worst specimen yet examined in a mine	18·2700

Dr. Smith goes on to say (p. 9):—'Some people will probably inquire why we should give so much attention to such minute quantities—between 20·980 and 20·999—thinking these small differences can in no way affect us. A little more or less oxygen might not affect us, but, supposing its place occupied by hurtful matter, we must not look on the amount as too small. Subtracting 0·980 from 0·999 we have a difference of 190 in a million. In a gallon of water there are 70,000 grains; let us put into it an impurity at the rate of 190 in 1·000,000: it amounts to 13·3 grains in a gallon. This amount would be considered enormous if it consisted of putrefying matter, or any organic matter usually found in waters. But we drink only a comparatively small quantity of water, and the whole 13 grains would not be swallowed in a day, whereas we take into our lungs from 1,000 to 2,000 gallons of air daily.'

[1] *Air and Rain: the Beginnings of a Chemical Climatology.* London: Longmans. 1872.

In addition to the two gases, oxygen and nitrogen, the atmosphere contains other constituents which are constantly present in it, but in varying, and always very minute, proportions. The most important of these are carbonic acid and aqueous vapour.

The first-named gas is notoriously deleterious to animal health. It is the product of the complete oxidation of carbon, as effected in combustion, and therefore in respiration, which is a slow combustion, and it is absorbed by plants under the influence of sunlight.

Its normal amount may be taken as 0·03 per cent. An average of several specimens of London air, taken in November 1869, gave Dr. Smith 0·04394 per cent., while examples taken in the tunnels of the Metropolitan Railway, in the same year, gave 0·1452. Dr. Smith, again, says:[1]—'We all avoid an atmosphere containing 0·1 of carbonic acid in crowded rooms; and the experience of civilised men is that it is not only odious but unwholesome. When people speak of good ventilation in dwelling-houses they mean, without knowing it, air with less than 0·07 of carbonic acid.'

The proportions of oxygen and carbonic acid, however, are not subjects of ordinary meteorological inquiry; but it is very different with the aqueous vapour, which plays a most important part in all meteorological changes, and the determination and consideration of which will form the subject of special chapters.

As to ozone, which is known to exist in the atmosphere and which is only oxygen in what chemists call an allotropic condition, its determination is deemed by

[1] *Air and Rain*, p. 56.

some authorities to be of very high importance in connection with sanitary meteorology; but the assertion may safely be made that as yet no thoroughly unquestionable mode has been proposed for its detection and quantitative determination. To this subject, however, we shall return subsequently (Chapter X. p. 196).

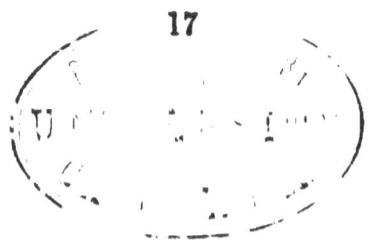

CHAPTER III.

TEMPERATURE.

THE determination of the temperature of the air is, without doubt, the most important of meteorological observations, for as Professor Angus Smith has pointed out, in his work already quoted, 'Heat is a more pressing want than even pure air.' If, however, the observation is important, and is apparently very easily made, it is yet almost the most uncertain of all, if great accuracy be required, owing to the nearly total impossibility of securing a perfectly unexceptionable exposure for the thermometers. The indications of these instruments will vary more or less, with the pattern and material of the stand in which they are placed, with their height above the ground, their proximity to trees or buildings, and even with the nature of the soil and the character of the vegetation which covers it.

We must commence with some notice of the thermometer itself, of which instrument a most interesting history has been published by M. Renou in Paris,[1] and we shall describe some of the principal forms of thermometers at present in use. An ordinary thermometer consists of a fine glass tube with a bulb blown on one end, and is partly filled with some liquid, usually mercury or spirit of wine. This liquid expands on being heated, and contracts again on being

[1] *Annuaire de la Société Météorologique de France*, 1876, p. 19.

cooled. When it expands it passes up along the tube. By the amount of this expansion the temperature is measured by means of a scale marked off on the tube. The mode of making and filling thermometers so as to exclude all air from the bulb and tube is given in text-books of physics, as in Balfour Stewart's 'Elementary Treatise on Heat' (1866). Other conditions being equal, the finer the tube in relation to the bulb the more sensitive will the instrument be, for the scale will be more open.

The actual inventor of the thermometer is unknown. At the end of the seventeenth century two forms of instruments for measuring temperature were recognised —Spirit thermometers with closed tubes (Florentine thermometers), and Air thermometers, in which the open end of the tube was plunged in water (Dutch thermometers). It is generally stated that Galileo was the real inventor of the first-named form of thermometer, but Robert Fludd, in a work published in 1638, expressly stated that he had found the instrument described in a manuscript dating more than fifty years back. This date is at least twenty years before Galileo went to Florence, and began to publish, for his first paper is dated *circa* 1609.

The great improvements in the instrument, which, in its simplest form, corresponds in principle exactly with the Florentine pattern, have been the substitution of mercury for spirit as the thermometric liquid, and the reference of the scale to the fixed points of boiling water and melting ice. All three of these improvements are placed by M. Renou to the credit of Englishmen, the first two having been suggested by Halley in the 'Philosophical Transactions' for 1693; while the third, the proposal to use the freezing-point of water as the starting-point of the scale, was made by Robert Hooke in 1667.

The advantages presented by the use of mercury as a thermometric liquid are the great distance between its melting-point and its boiling-point, amounting to about 700° on the Fahrenheit scale; the regularity of its expansion for the greater part of that range; its high conductivity and low specific heat,[1] which enable it to indicate more rapidly than other substances would the temperature to which it is for the time exposed.

The only conditions under which mercurial thermometers cannot be employed in meteorology are those of extreme cold, inasmuch as at the temperatures experienced in winter in very cold climates, such as those of Canada or Siberia, mercury freezes, or at least is so near its point of congelation that its expansion is not so regular as is the case between the limits of temperature met with in warmer climates.

Various plans have been proposed for the graduation of thermometers; of these three have come into extensive use: those of Fahrenheit, Celsius, and Réaumur. In Fahrenheit's thermometer, used at all British and American stations, the interval between the freezing and boiling-points is divided into 180 degrees, the freezing-point being 32° and the boiling-point 212°.

In the Centigrade thermometer, invented by Celsius, a Swede, and in almost universal use on the continent of Europe, the same interval is divided into 100 degrees, the freezing-point being 0°, and the boiling-point 100°.

[1] *Conductivity* is the rapidity with which heat is communicated from particle to particle of a substance. *Specific heat* is the amount of heat required to raise 1 lb. of a substance one degree, in terms of that necessary to raise 1 lb. of water one degree.

In Réaumur's thermometer, formerly in use in Germany and Russia, the same interval is divided into 80 parts, the freezing-point being 0°, and the boiling-point 80°.

In both the Centigrade and Réaumur's scales the degrees below the freezing-point are marked with a *minus* sign (−), and hence arises the convenient, but slovenly, way of speaking of ten or twenty degrees of frost, as the case may be, instead of −10° or −20°.

In Fahrenheit's scale, as the freezing-point is 32°, it is found that the minus sign is comparatively rarely used, at least in these islands, as the temperature hardly ever sinks as much as 32 degrees below the freezing point.

Of these scales, Réaumur's, which was formerly very common, is now rapidly falling out of use, owing to the very general adoption of the Centigrade system.

Inasmuch, however, as all three scales are in existence and occasionally met with, it is well to know the following rules for converting readings according to one into readings according to either of the others.

To convert Fahrenheit readings to Centigrade, subtract 32 and multiply the remainder by $\frac{5}{9}$; *e.g.* 68° F when reduced becomes (68−32) $\frac{5}{9}$ = 20° C.

To convert Fahrenheit readings to Réaumur, subtract 32 and multiply the remainder by $\frac{4}{9}$; *e.g.* 68° F when reduced becomes (68−32) $\frac{4}{9}$ = 16° R.

To convert Centigrade to Fahrenheit, multiply by $\frac{9}{5}$ and add 32.

To convert Réaumur to Fahrenheit, multiply by $\frac{9}{4}$ and add 32.

To convert Centigrade to Réaumur, multiply by $\frac{4}{5}$.

To convert Réaumur to Centigrade, multiply by $\frac{5}{4}$.

Tables for these conversions will be found in Appendix I., but in this work Fahrenheit's scale will be exclusively employed.

It will now be interesting to describe the graduation of the instrument more minutely. In the first place the freezing-point is the temperature of pure water with ice melting in it, or of melting snow. It is necessary to state this, because the temperature at which water freezes is not constant, as will be learnt from text-books on physics, and under certain circumstances water may be cooled several degrees below the freezing-point without the formation of ice.

The boiling-point requires further definition. The temperature at which water boils depends on the pressure of the atmosphere at the time, as the less is that pressure the easier is it for the vapour to liberate itself from the liquid, and accordingly the lower is the temperature at which ebullition occurs.

Hence we see how the observation of the boiling-point of water may be employed to measure atmospherical pressure. As we shall subsequently learn (p. 65) that the pressure of the atmosphere is reduced as we ascend above the level of the sea, and in fact depends on the elevation we have reached, we can see that any method of determining the pressure at any station may be used to ascertain the height of that station. The observation of the temperature at which pure water boils is a method, though an indirect one, of ascertaining the pressure, and as a boiling-point thermometer, with a lamp attached, is an easily portable apparatus, the method is a convenient one for estimating the heights of mountains. Tables have been calculated to show the height corresponding to any observed temperature at which water boils. The proces is a fairly accurate one.

The standard pressure which is assumed for th boiling-point in Fahrenheit's thermometer is 29·90: ins. in the latitude of London; and that adopted fo the boiling-point on the Centigrade scale is 760 milli mètres in the latitude of Paris.[1] As this latter pressur corresponds to 29·922 ins. it is evident that 212° F. an 100° C. do not indicate *precisely* the same temperature although the approximation is sufficiently close fo ordinary purposes.

To return then to Fahrenheit's graduation; h divided the distance between the freezing and boilin points into 180 parts or degrees, corresponding to th number of degrees in a semicircle, and he assumed a the zero of his scale the temperature which resulte from the mixture of snow and salt. This was 3 degrees of his scale below the freezing-point of pur water. In these thermometers, therefore, the minu sign (—) is only used for temperatures below 0° F.

In addition to the advantages presented by th Fahrenheit's scale, owing to the comparative rarity c the use of the negative sign, the small size of its degree renders it very convenient.

The types of thermometers which are principall employed in meteorological observations are ordinar thermometers; maximum thermometers; minimur thermometers; and solar radiation thermometer which will be described in a subsequent chapter), o which the three last-named types are ordinary ther

[1] It is necessary to state the latitude, as the action of gravity depenc in the latitude, and so does consequently the weight of the mercury require to produce the requisite pressure.

mometers arranged so as to be self-registering, that is they are provided with an index to mark the extreme of temperature which has been registered since the instrument has last been set. In addition to the foregoing, some observatories are provided with automatically self-recording thermometers, which furnish a continuous record of temperature, independently of eye observations.

Ordinary Thermometers.—An ordinary thermometer is mercurial, and it should have a small bulb, cylindrical by preference, but for practical purposes spherical bulbs will suit well enough. The tube should be fine enough in proportion to the size of the bulb to allow of an extended scale, and the graduation should be carried out at least from the lowest to the highest temperature to be expected to occur at the station. Thus, for these islands, the range from $-10°$ to $100°$ will suffice for ordinary shade temperatures, as the temperature will hardly ever sink below $0°$ or rise to $100°$, under the ordinary conditions of observation.

Every thermometer should be verified at some recognised establishment, such as Kew Observatory. A Kew certificate will show the instrumental errors at every ten degrees from $32°$ to $92°$, or for a greater range in special cases. The rule laid down by the Meteorological Office is that no thermometer should be used which has at any one point a larger error than $0°·3$, or in which any space of ten degrees is more than $0°·3$ wrong. Thus, if the correction at $52°$ is $+0°·2$ and at $62°$ is $-0°·2$, the instrument should be rejected. That these limits of accuracy are not too stringent is shown by the fact that several of the

London makers annually turn out hundreds of instruments amply fulfilling the conditions stated.

There is, however, a source of inaccuracy in thermometers against which special precautions must be taken. This is called the 'displacement of zero.' When glass has been fused, it does not at once return to the precise conditions of density, &c., which it exhibited before being heated. The bulb has a tendency to contract, and this action of course causes it to hold less mercury, and, by forcing a portion of its contents into the tube, makes the instrument read too high. The progress of this contraction is not perfectly regular, and it may last over a considerable period. The best method of obviating it is for the maker to anneal the tubes as soon as they are filled and sealed, and then to lay them aside, but not to graduate them until they have been lying by for several years.

It is, however, comparatively easy to discover the existence of this displacement of zero in an instrument by testing it in ice or snow, and to correct it if it exists. Each winter, when snow is on the ground, the thermometer should be placed in melting snow and entirely covered up, with the exception of a small portion of the scale just above the freezing-point. Of course pounded ice would do as well as snow, but the latter is more easily manipulated. As soon as the mercury has taken the temperature of the melting snow, which will be shown by the reading remaining constant, the precise position of the end of the column should be noted. Any difference found between this reading and 32° should be entered as the correction for zero displacement. This correction is always subtractive, and should be deducted from all the readings of the instrument, as the accuracy of graduation of the tube is entirely independent of the capacity of the bulb; so that any error in the estimation of the latter will affect equally the whole range of the scale.

Self-registering Thermometers.—Self-registering thermometers have already been defined to be instruments furnished with some contrivance to mark the highest or the lowest temperature to which they have been exposed during a given interval of time. Such thermometers are usually read once a day at a definite hour, and then set to agree with the temperature at the time.

FIG. 1.

The oldest form of self-registering thermometer is Six's, invented in 1781. It combines the registration of the maximum and minimum temperatures in one instrument. It is very commonly employed in greenhouses, &c., but is never used for scientific observations, except for ascertaining the temperature of the deep sea. It is a spirit thermometer, the tube of which is bent parallel to itself, and a bulb is blown on each end, one bulb being larger than the other (fig. 1). The two bulbs are filled with spirit, except that a bubble of air is left in the smaller one. The bend of the tube is occupied by a plug of mercury. The registration is effected by means of two indices, consisting of steel pins, sealed in glass tubes, with hairs attached to keep them in any position they may have reached by being pushed by the mercury plug, or by the action of

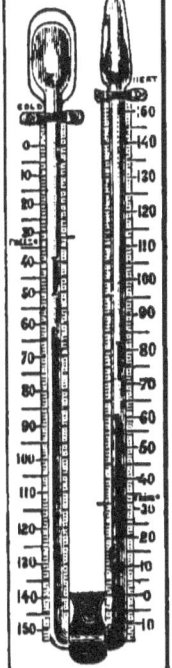

Six's Thermometer

the magnet, which must be employed to set the instrument.

The setting consists in bringing down the indices, by means of the magnet, to rest on the top of the mercury plug, the maximum on the side of the smaller, the minimum on that of the larger bulb. When the temperature rises, the spirit in the large bulb expands, and pushes the plug and index before it, the latter resting at the highest point reached. When the temperature falls, the spirit contracts, and the pressure of the air-bubble in the small bulb drives the plug back, and moves the minimum index in its turn as soon as the temperature falls below that at which the instrument was set. This index then marks the lowest point reached.

The instrument is apparently very complete, but it has serious defects. It must always be kept in a vertical position, or else the spirit may creep past the mercury at the bend of the tube, and then the two ends of the mercury will indicate different temperatures, and the instrument must be sent to an optician to be set right. Again, the mercury has a tendency to push itself beyond the ends of the indices, so that small quantities are retained by them. By this means the length of the mercurial column is reduced, and the instrument rendered incorrect. Practically, the maximum and minimum thermometers are now always distinct instruments, and Six's thermometers are never used at well-appointed stations.

Maximum Thermometers.— Of maximum thermometers the two commonest patterns are Phillips's and Negretti's, and there is not very much to choose between them. In Phillips's thermometers the index is formed by a small portion of the mercurial column separated from the main thread by a bubble of air. This separated portion is pushed on before the column when it advances, but does not return with it when it recedes, so that the index rests at the extreme position which the column has reached, and the end furthest from the bulb registers the highest temperature which has been attained. Thus the reading of the instrument shown in fig. 2 is 78°.

The chief objection to this form of thermometer is that the bubble of air is likely to become displaced by passing back into the bulb if the temperature falls very

Fig. 2.

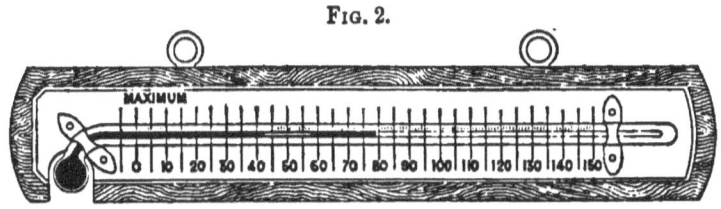

Phillips's Maximum Thermometer

low. When this happens the instrument loses its registering properties, and becomes an ordinary thermometer.

The plan of Negretti's thermometer is simple, and the instrument is rather less likely to get out of order than the preceding, for which reason it may be considered preferable for ordinary use. The registration is effected by the mercurial column itself in the following manner. The bore of the tube close to the bulb is reduced in section in such a way that while the expansion of the mercury is sufficient to force the thread of liquid past the obstruction, the cohesion of the metal is insufficient to draw it back again when the temperature falls.

Accordingly the length of the thread of mercury

above the contraction measures the highest temperature to which the instrument has been exposed since it was last set. The thermometer should be held with the bulb slightly inclined downwards, before reading, to allow the separated portion of the column to flow gently back to the contraction.

Both these thermometers are set in the same way. The instrument is taken in the hand and swung briskly, bulb downwards. This has the effect in Phillips's instrument of causing the index to come down till stopped by the air-bubble resting on the continuous portion of the thread. In Negretti's instrument the centrifugal force will enable the separated portion of the column to pass the contraction and enter the bulb, until the latter is quite full and the instrument indicates the actual temperature of the air at the time.

The hands must be kept away from the bulb when setting a maximum thermometer, for otherwise it will be almost impossible to set the instrument so as to show the true temperature of the air, especially in winter.

In making Negretti's thermometer, great care is requisite to ensure that the contraction shall be neither too great nor too slight. If the former defect exists it will be found difficult to set the instrument, while if the tube is not contracted enough the indications may be inexact, as some of the mercury may, by mechanical adhesion, be drawn back past the contraction when the temperature falls.

Precautions are required in placing any registering thermometer on its stand. It ought to hang exactly horizontal; but some instruments, owing to slight defects of construction, require to be sloped *very slightly* to ensure their acting properly.

Minimum Thermometers.— Of minimum thermometers there are two classes, according as the instrument is filled with mercury or spirit. Of the former class Casella's is that best known. It is an exceedingly ingenious and beautiful instrument, but requires so much care in its manipulation that it is practically but seldom used.

Rutherford's pattern is that all but universally adopted for spirit minimum thermometers. In it the index is entirely enveloped in the liquid and moves with a little difficulty in the tube. The action is as follows :—The index is allowed to run down to the end

Fig. 3.

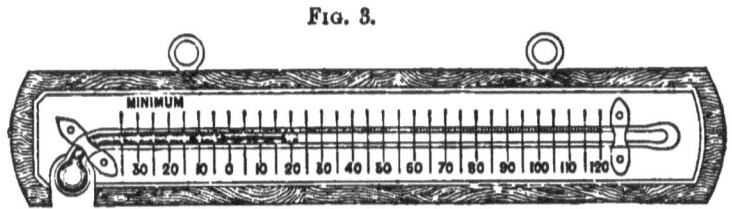

Rutherford's Minimum Thermometer.

of the column by sloping the thermometer with the bulb uppermost. When so set it is placed in a horizontal position. If the temperature rises the spirit will flow past the index without disturbing it, but if it falls below the point which the thermometer marked when set, the force of capillary attraction between the spirit and the index is sufficient to preclude the index being left dry, and accordingly this is drawn back with the spirit until the lowest temperature is reached, its upper end being always flush with the end of the column as that recedes. The index will thus mark the greatest degree of cold which has occurred, inasmuch as it will remain unmoved when the spirit advances again, owing to any rise of temperature. The reading of the instrument shown in fig. 3 is accordingly 17°.

The lowest temperature since last setting is of course taken from the position of the index, but the column of spirit itself will mark the temperature at any time, and if the two readings are identical it is a sign that the temperature is at its lowest and probably falling.

Rutherford's thermometer is liable to a serious defect: a portion of the spirit may evaporate from the end of the column and condense in the upper part of the tube. This arises from the presence of air in the spirit used for filling the instrument. The quantity of liquid so condensed sometimes amounts to six or eight degrees, and of course the length of the continuous column, by which the temperature is recorded, is curtailed to that extent, and the readings lowered proportionably. It is probably to this source of error in spirit thermometers that most of the extraordinary discrepancies in reports of severe cold are to be attributed. In fact, it may safely be said that more than half the ordinary spirit thermometers in use in gardens are to a greater or less extent affected by this error. If a spirit thermometer reads lower than a correct mercurial thermometer placed beside it, there is reason to suspect the presence of the defect just mentioned.

The spirit is also liable to become broken into several detached portions, especially if the instrument is being transported from place to place; or the index may be shaken entirely out of the spirit into the upper part of the tube. In all these cases the thermometer should be swung briskly to and fro several times, holding it bulb downwards, until all the liquid which may have been visible at the upper end of the tube shall have been dislodged and the index brought down to the bulb end of the bore.

The thermometer should then be placed in an upright position, bulb downwards, and left there for about an hour. This treatment will usually have the effect of restoring to the instrument its correctness of indication. If, however, it should fail to do so, the thermometer should be held in the right hand about a third of the way up from the bulb, and the upper end of the tube gently tapped on the palm of the hand, which will usually dislodge the spirit from that place, and cause it to unite with the main column.

We have already mentioned the high conductivity and low specific heat of mercury as qualities which render it the most useful thermometric liquid, inasmuch as they render it more sensitive to changes of temperature than spirit. The sluggishness on the part of the latter liquid is obviated by the adoption of a different mode of construction of the bulb. Of all the forms of bulb, the sphere is the worst for the sensibility of the thermometer, as that form presents the least amount of surface for a given mass of liquid. The cylinder is far preferable to the sphere, but for spirit minimum thermometers this form will hardly suffice and several plans have been devised to render the instrument sufficiently sensitive. One of these is to make the reservoir forked in shape, another is to form it out of two pieces of glass tubing of different diameters, placed one inside the other and fused together at each end, so that when the instrument is filled the liquid forms a thin sheet, of which no portion is more than a tenth of an inch in thickness. Any alteration in temperature will by this means be simultaneously communicated to a large surface of spirit, and will speedily produce its effect on the whole mass, so that the indications will be as prompt as those of a mercurial instrument.

Self-recording Thermometers.—All the thermometers which we have been considering hitherto require to be read at frequent intervals; but as such a practice is very troublesome, and at best cannot keep account of all the changes of temperature which occur, various methods have been adopted to obtain a record of temperature independently of the constant attendance of the observer.

Record at frequent intervals can be obtained by the use of automatic instruments in which electricity is employed, and which are termed electrical *thermographs*. In Theorell's and van Rysselberghe's, which are amongst the best of these, the thermometer differs from ordinary thermometers in that the tube is open at the upper end, and a wire is introduced into it which, by a clockwork arrangement, on the plan suggested by Sir C. Wheatstone, is caused to descend at regular intervals until it touches the surface of the mercury. As soon as contact is established, an electric current is set up and a record is obtained. The wire is then raised again and the contact is broken. The record is furnished by a mechanical contrivance which, whenever contact is made, pricks a hole in paper, on which the thermometer scale is marked, at the point corresponding to the height of the mercurial column at the time.

A continuous record is furnished by the photographic thermograph, adopted by the Meteorological Committee at their observatories. In this instrument a bubble of air is introduced into the column of mercury, which moves up and down with the temperature, for the bore of the tube is larger than in Phillips's maximum thermometer, in which the separated portion does not return towards the bulb when the temperature begins to fall. A lamp is placed before the instrument, and a photograph of the space occupied by the air-bubble is taken on prepared paper, stretched on a drum, and caused to revolve at a fixed rate.

Another form of photographic thermograph, which is that in use at Greenwich, furnishes a photograph of the open space in the thermometer tube, above the mercury, so that the length of the photographic impression varies with the thermometric reading.

Thermometer Exposure.—When we have got our thermometers, the next question is how to place them so as to give a correct indication of the temperature of the atmosphere. This is a problem of the greatest difficulty. In the first place, it is evident that an instrument intended to give the temperature of the air must be protected against radiation, so that it shall not be possible for it to receive appreciable heat from bodies hotter than the air or to give out appreciable heat to bodies cooler than the air.

On the other hand, the air must have perfectly unrestricted access to the bulb, and that air should be the free air of an open space, not the confined air of a narrow courtyard. This last condition is difficult of fulfilment in certain places, for it virtually implies the assertion that no observation of temperature taken in a town is of any value at all. Moreover, if we demand that the screen shall always be erected at a distance from any building, we at once condemn not only the exposure of all thermographs, but the observations taken at most of the stations on the continent of Europe, where the use of screens at windows is very generally adopted.

Practically, a screen standing free over an open grass-plot should be employed wherever possible; but if the best exposure available be a wall, care should be taken that it is a garden wall—not the wall of a house which may be heated by fires inside—and that the screen hangs at a distance of some inches from the wall so as to admit of the free passage of the air behind it. The screen should face north in the northern hemisphere and be sheltered from the sun at all hours, but it must be exposed to a free circulation of air.

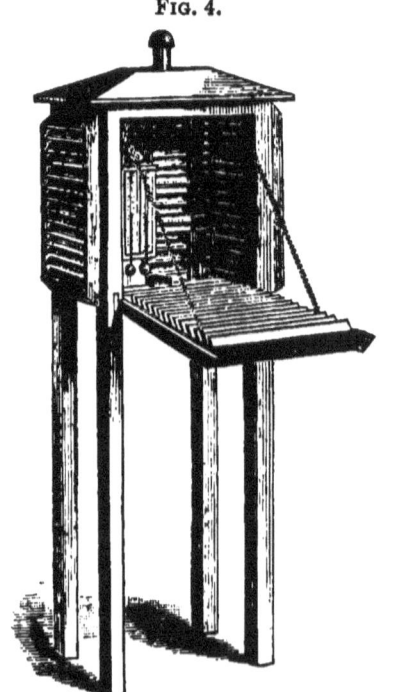

Fig. 4.

Stevenson's Thermometer Screen.

As to the pattern of the screen, that most usually employed in this country has been devised by Mr. Thomas Stevenson, C.E., and is shown in fig. 4. It is made of wood, double louvred, and is erected on legs, so that the bulbs of the thermometers when suspended in it shall be four feet above the ground. Its external dimensions are—length, 23 in.; breadth, 14 in.; height, 18 in. This and all other screens should be painted white. The objections to Stevenson's screen are that it is rather small, and that the louvres are too close together to allow of a free access of the external air to the bulbs.

Other forms of screen were formerly much in vogue, such as Glaisher's or James's. These are perfectly open on the side away from the sun, and their main defects are that they require to be shifted once or twice a-day to prevent the sun's rays falling on the

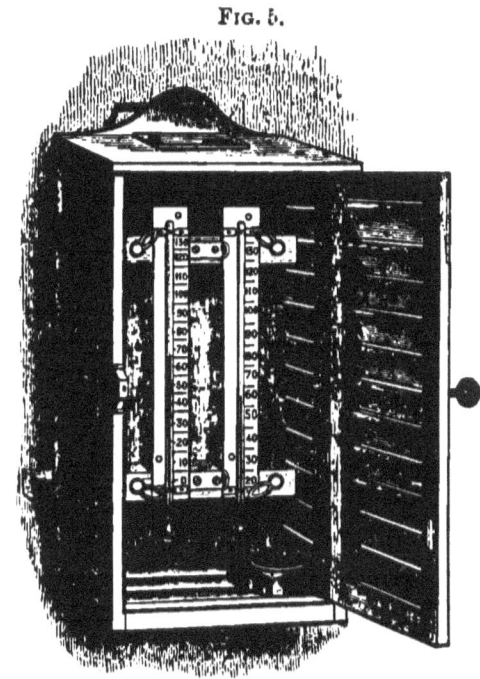

Fig. 5.

Thermometer Screen used on board Ship.

thermometers, and that in time of rain with a northerly wind it is impossible to keep the instruments dry. Finally, these screens do not protect the thermometers against radiation. For the last-mentioned reason the maximum and minimum readings recorded in such screens are frequently higher and lower respectively than those registered in a closed screen.

Of wall screens, that used on board ship, fig. 5 (p. 35), is very good, and its construction is extremely simple, but when it is set up on a wall facing north, so as to keep it in the shade, some additional protection is usually required on the east or west sides or on both, to shelter it from the rays of the sun at its rising and setting in summer. It should be placed so that the thermometer bulbs should be at a height of four feet above the ground, and its back should be distant two or three inches from the wall.

Sling Thermometer.—In order to meet the difficulty of exposure, Arago, in 1830, proposed to determine the temperature by means of a thermometer tied to a string and swung round the head. An observation taken under such conditions gives a very close approximation to the shade temperature at all times, and it is a remarkable fact that the reading is scarcely affected by the accident of its being taken in sunshine or in shade. It is obvious that such a method of observing obviates all the difficulties about the free circulation of the air round the thermometers which have been mentioned in connection with Stevenson's screen. It is, however, only suited for isolated observations.

We now come to the arithmetical treatment of the observations. The results may be considered in two directions: as to the changes of temperature firstly *in time*, and secondly *in space*.

The question of the changes of temperature *in time* brings with it the whole subject of its diurnal and annual range: what are called its Periodic Variations. Under the heading of the changes of temperature *in space*, we have to deal with the great subject of climatology, and to consider not only how the average temperature varies from place to place, but how the curves of annual and diurnal range are modified with change of geographical position. These latter subjects will be dealt with in Chapters XII. and XVII.

Mean Temperature.—When we obtain by photographic or mechanical means a continuous record of thermometric readings for any length of time, we find that the temperature is constantly changing and that,

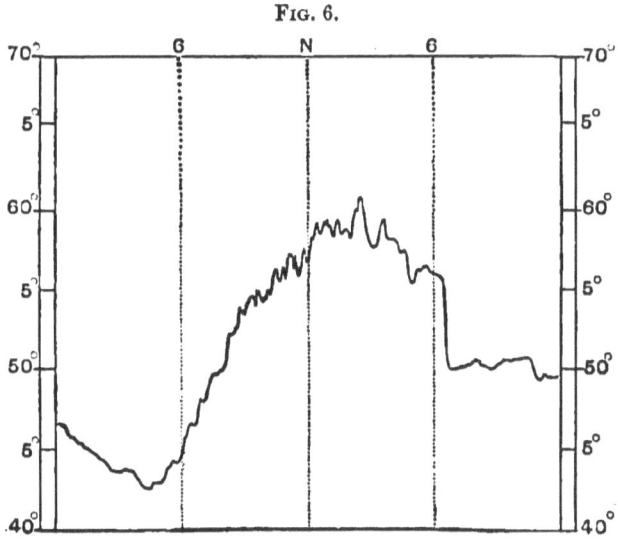

Thermogram, Kew Observatory, August 3, 1877.

in fact, on a hot summer's day, the curve is so constantly showing sudden variations that it is often only a chance whether or not two consecutive readings taken within a short interval of each other will accord within a degree or so.

The diagram (fig. 6) shows the continuous record of temperature taken at Kew Observatory during a summer day.

38 ELEMENTARY METEOROLOGY.

The average temperature for the day ought to be obtained by *integrating*, or determining the area enclosed by, this irregular curve;[1] but as it is very difficult to integrate such curves as that given above, the thermograms, as such curves are called, are measured at every hour, and it has been generally agreed to take the average of 24 such measurements, or of 24 hourly observations, if attainable, as the true mean of the day.

It is, however, impossible to obtain continuous records from many stations, and far too troublesome for any ordinary station to maintain hourly observations, and accordingly calculations have been made to find the degree of approximation to the mean of the 24 hourly readings which is attainable by any combination of less frequent observations. The following figures

Month	9 a.m.	½ (9 a.m. + 9 p.m.)	½ (max. + min.)	⅓ (6 a.m. + 2 p.m. + 10 p.m.)	⅓ (7 a.m. + 1 p.m. + 9 p.m.)
January . .	1·0	0·8	0·3	− 0·3	− 0·3
February . .	1·0	0·9	− 0·4	− 0·2	− 0·2
March . .	− 0·1	0·7	− 0·7	0·1	− 0·1
April . .	− 1·5	0·4	− 1·1	0·4	− 0·5
May . . .	− 2·4	0·2	− 1·2	0·3	− 0·6
June . .	− 2·5	0·2	− 0·8	0·3	− 0·9
July . . .	− 2·2	0·3	− 1·0	0·4	− 0·8
August . .	− 1·9	0·4	− 1·2	0·5	− 0·5
September .	− 1·5	0·5	− 1·1	0·2	− 0·3
October . .	− 0·4	0·6	− 0·8	− 0·1	− 0·3
November .	0·2	0·8	0·1	− 0·2	− 0·3
December . .	0·8	0·6	0·3	− 0·3	− 0·3
Year . .	− 0·7	0·5	0·6	0·1	− 0·4

[1] It is obvious that the area enclosed by any curve like that shown in fig. 6, of which the *abscissæ*, measured along the base, represent time, and the *ordinates* represent values of any element, may be represented by a rectangle on the same base, and of which the adjacent side is the ordinate representing the mean value of the element for the time.

show, for Greenwich, the corrections to be applied to the ordinary combinations of hours of observation, in use in Europe, in order to obtain true daily means. The data on which this table has been based are given in Table 1, p. 38, of the 'Reduction of the Greenwich Meteorological Observations,' and extend to the range of years 1849–1868. The correction for the maximum and minimum readings is from a paper by Mr. W. Marriott.[1]

In the table the figures are of course degrees and tenths Fahrenheit.

The values marked with a *minus* sign (−) are higher, those without a sign are lower, than the true daily mean for the month.

These figures are, however, strictly speaking, only applicable to the locality of Greenwich, and require modification before they can be employed for observations taken elsewhere.

This brings us to a very important subject. It is obvious that as the average of each combination of hours of observation differs by a definite quantity from the average of the day, so the average of any single observation taken regularly at a fixed hour, as for instance that for 9 A.M. in the preceding table, must differ by an ascertainable amount from the average of the day. Once this amount was known, it would therefore be only necessary to observe once a day in order to determine the temperature of the station. But the corrections deduced from any one station are not applicable to other stations differently situated as to their geographical positions and their heights above the sea-level. The mean temperature deduced for a

[1] Q. J. Met. Soc. vol. iii. p. 402.

number of stations by such a method by means of a table of corrections determined for one of them will differ appreciably from the mean temperatures obtained from a greater number of observations taken at the stations themselves.

The subjoined curve (fig. 7) shows the range of mean temperature at Greenwich during the day on the average of the whole year, as derived from the twenty years' observations, 1849–1868. It will be seen from it

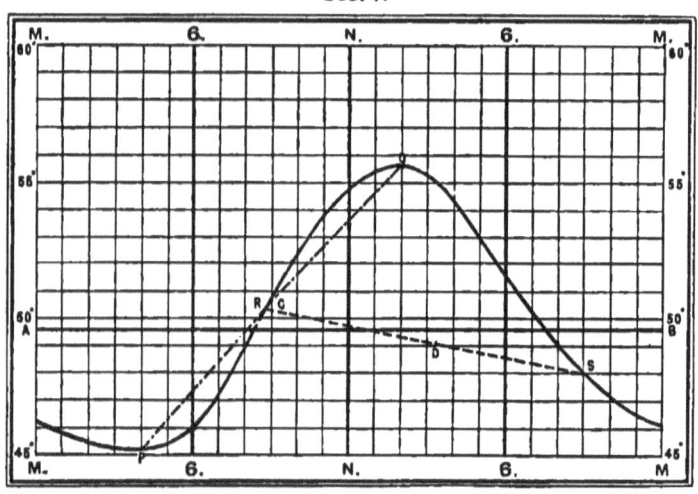

Fig. 7.

Diurnal Range of Temperature at Greenwich.

that the highest temperature occurs at 2 P.M., and the lowest at 4 A.M.

The course of this curve varies from month to month, with the season, and even from day to day, with the weather. As, however, the appearance of such a curve as this forms one of the most important characteristics of the climate of a place, the subject of diurnal range will be more fitly treated of when we come to deal with the distribution of temperature in Chapter XII.

In the foregoing diagram, the line A B represents the mean temperature of the year, 49°·7.

It is a fortunate thing that reasonably close approximations to the mean temperature of a day are obtained by taking the average of the observations at the homonymous hours of 9 A.M. and 9 P.M., as shown by the table on p. 38. The point D (49°·1) in the diagram shows the position of that average for Greenwich, being the middle point of the line R S, joining the 9 A.M. and 9 P.M. values. This combination is reasonably convenient for British observers, and has been adopted generally in this country, by the Meteorological Office and the Meteorological Society, for use at all stations of the Second Order.[1]

On the Continent, at the corresponding stations, more frequent observations are taken, and various combinations of hours have been adopted, usually three in number. The Russian system demands observations at 7 A.M., 1 P.M., and 9 P.M.; the Austrian at 6 A.M., 2 P.M., and 10 P.M.

The monthly values of the corrections for each of these combinations are given in the table (p. 38), and from these it appears that for Greenwich the series 6.2.10 gives the most accurate results.

[1] At the first International Meteorological Congress at Vienna, 1873, the following definitions of the different classes of stations were adopted:—

Stations of the First Order are observatories, in which meteorological observations are conducted on an extensive scale, either by the use of self-recording instruments, or by taking hourly readings.

Stations of the Second Order are those where complete regular observations of the usual meteorological elements, viz. pressure, temperature, humidity, wind, cloud, rain, &c., are conducted.

Stations of the Third Order are those where only a portion of these elements are observed.

However, the simplest method of determining the mean temperature is to take the average of the maximum and minimum temperatures for the twenty-four hours. This is the point o in the diagram, the middle point of the line P Q joining the highest and lowest readings.

The great convenience of this mode of determining the mean temperature arises from the fact that the maximum and minimum readings are registered automatically, so that the instruments need only be visited once a day.

A good deal depends on the hour at which the maximum and minimum thermometers should be set, and the period now fixed for this operation is 9 P.M., the latest observing hour of the day. The reason of this arrangement is easily understood. Theoretically these instruments ought to be set at midnight, so as to show respectively the highest and lowest temperatures reached during the civil day. This, however, is not convenient, and so formerly the plan was generally followed of setting them at the earliest observing hour in the morning. Such a practice, however, produced much uncertainty; for as, under ordinary circumstances, the maximum temperature recorded in the morning will have occurred during the afternoon of the previous day, while the minimum will have taken place about sunrise on the morning on which the observation is read off, the two readings will really have belonged to two different civil days. This gave rise to the absurd practice of assuming that the two readings always referred to different days, whereas, as a matter of fact, and in winter time especially, the maximum or minimum may occur at any period of the day, the extremes of temperature in cloudy weather being regulated mainly by the changes of wind.

If it were convenient to read the instruments at midnight no uncertainty could possibly exist, as the extreme temperatures recorded must have occurred during the civil day, and the adoption of 9 P.M. renders it at least far less likely that any doubt should arise as to the day to which a given temperature belongs than would be the case if a morning hour were selected.

Having calculated the mean temperature of the day by any method, the next problem is to determine the means for longer periods, and ultimately to arrive at the mean temperature of the place.

The most obvious period, next to a day, is a week, but as the weeks do not begin on the same days in consecutive years, we cannot easily state what is the true mean temperature of, say, the sixteenth week in any year.

The only really logical subdivision of the year is that into 73 periods of 5 days each, for the number 365 is only divisible by 5 and 73, and a great deal of good work has been done with these five-day means, which are strictly the means of equal and corresponding intervals, except in leap year, when the time from February 25 to March 1 embraces six days.

There is only one inconvenience in the use of five-day means, and this arises from the fact that the five-day intervals do not correspond with the monthly periods, while common usage always considers the month as the most natural subdivision of the year. Monthly means are, therefore, always employed for ordinary purposes, but in many respects this is an unsatisfactory mode of dealing with the question of the march of temperature during the year.

The months are unequal in length, so that, *e.g.*, the mean for February is not strictly comparable with that for either January or March, which are each of them the average of thirty-one days' observations, not of twenty-eight. Moreover, the absolute mean monthly temperatures do not fall on the middle days of the months—but occur on other days, so that the curve drawn from ordinates representing the twelve monthly means does not coincide closely with that drawn from the 73 five-day, and much less with that drawn from the 365 daily, means.

However, in all such questions something must be conceded to common usage, and so it has been agreed among meteorologists to calculate means for civil days beginning at midnight, instead of for astronomical days beginning at noon; and for civil months; while as the mean of the year, the average of the twelve monthly is to be taken instead of the average of the 365 daily means.

These mean monthly and annual results should then be combined in periods of five years called *lustra*, the periods being so chosen that the year with which they end shall be a multiple of five. Thus the average should be taken of the five years 1871–5 inclusive or 1876–80 inclusive, but *not* of 1872–6 or 1873–7.

It must be remembered, when we are speaking of mean temperatures, that the annual mean of itself gives us no information at all as to the extent to which the individual monthly means respectively exceed or fall short of the yearly average.

To give an example: the mean temperature of

Dublin is given as 49°·3, and that of New Haven, in Connecticut, as 49°·2, yet the climates of these two places offer strong contrasts to each other. The temperature of January in Dublin is 40°·6, and in New Haven 26°·6, while the temperature of July in Dublin is 59°·7 and in New Haven 71°·6, so that the difference between the hottest and coldest months is only 19°·1 in Dublin as compared with 45° in New Haven.

It need hardly be said that, in general, equable climates are more favourable to human life than those which are subject to great diurnal or annual fluctuations, and in this respect the United Kingdom possesses rare advantages, as we shall see when we come to deal with the question of Climate (Chapter XVII.).

It will now be interesting to see how the changes of temperature of which we have been speaking are brought about. The only source of heat which we practically need take into consideration is the sun. Why he gives out so much heat, how long he has done so, and will continue to do so, are questions which are among the most interesting which cosmical science presents to us, but they do not come under the head of meteorology. It is enough for us to know that we feel more heat when the sun is shining than when he is not. The sun's heat-rays, like those of light, come to us in straight lines, if we neglect the effect of atmospheric refraction, and strike on the hemisphere which faces the sun. If the earth and sun were both fixed, half of the globe would be in sunshine and the other half in shade, and the hottest part would be that nearest the sun. But the earth turns on its axis, and at certain times of the year, the equinoxes, the line joining the centres of the earth and sun is perpendicular to the earth's axis. At these times therefore every portion of the earth's surface is successively exposed to the rays of the sun.

This is the case only twice a year, and as the earth moves in her orbit the angle of inclination to the axis of the earth of the line joining the centres of the sun and the earth changes through $23\frac{1}{2}°$ on each side of its equinoctial value, as explained in Chapter II.

This apparent motion of the sun in the ecliptic is the key to all our seasonal changes, which would be simple enough were the surface of the earth uniform in character. This is, however, far from being the case: part of it is always solid, part always liquid, and part solid and liquid by turns, being that portion of the globe where water remains frozen during the winter. We shall return to this subject in Chapter XII. p. 219.

The effects of these several conditions of the surface in modifying the influence of heat are most important. Water is heated with great difficulty, and gives up the heat, which it has once acquired, with equal reluctance. Land is heated easily, and has a relatively high radiating power. Ice is heated slowly, and the effect of heat on it while it is being melted is scarcely appreciable to the thermometer, owing to the absorption of heat, which is rendered 'latent' in the process of change of state from solid to liquid.

We see, therefore, that the temperature of land rises quickly during the daytime and falls as rapidly during the night, while water is much less sensitive to the influences of absorption and radiation of heat.

We shall return to this subject in the next chapter, when treating of Radiation; but, to show the height to which the temperature of a dry sandy soil may rise under the action of the sun's rays, it may be stated that Sir John Herschel observed a superficial temperature of 159° in such soil at the Cape of Good Hope; and Sturt says that, on the Macquarie River in New South Wales, 'the ground was almost a molten surface, and if a match fell upon it it immediately ignited.'

The theory of the movable equilibrium of heat is that all bodies tend to place themselves in what is called thermic equilibrium with surrounding bodies. Heat is passing in all directions between all the objects, say, in a room. If the heat received by any object exceeds that given out by it its temperature rises; if the contrary is the case its temperature falls. It is only when the gain and the loss are exactly balanced that the temperature remains constant.

These considerations enable us to trace the course of events as regards heat on a sunny day. The amount of heat received from the sun increases hour by hour from his rising until noon, and then decreases again till sunset, while all night long we receive no heat at all from him. The radiating power increases almost *pari passu* with the increasing heat, but cannot quite keep pace with it, and so the day grows warmer as it wears on. The heat received begins to decrease when noon is passed, but the amount given off does not equal that received for about two hours, and, accordingly, the hottest part of the day is about 2 P.M. The coldest is just before sunrise, because then the influence of solar heat has been withdrawn for the longest possible period, while the earth all the time has been radiating heat out into space.

If we apply a similar train of reasoning to the yearly period we shall understand how it is that the hottest month in these latitudes is not June, but July, and the coldest not December, but January. The common saying,

'As the day lengthens the cold strengthens,'

expresses this fact well.

The following diagram exhibits in a very concise and intelligible way the entire march of temperature

FIG. 8.

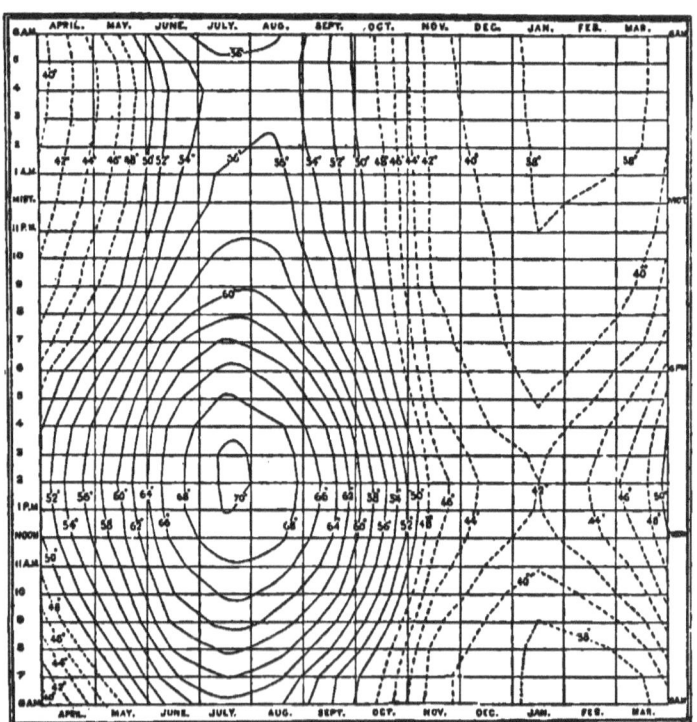

Chronoisothermal Diagram representing the Monthly Mean Temperature at Greenwich, for every hour of the day through the range of years 1849 to 1868.

during the year. It is taken from Table L. of the 'Reduction of the Greenwich Meteorological Observations' (1878), which is also reproduced in the 'Quarterly Journal of the Meteorological Society,' vol. iii., p. 400.

In this diagram, the principle of which was devised by M. Léon Lalanne, and the construction of which is explained in Appendix II., the horizontal lines are hours of the day, and the vertical lines months of the year. The curved lines are called *chrono—iso—thermals* (time—equal—warmth). Those for 50° and above are drawn full; those for degrees below 50° are broken lines.

The diagram shows that the highest mean temperature (70°) only occurs in the latter part of July, and between one o'clock and half-past three P.M. The lowest mean temperature (38°) is observed during the night time from about January 5 to March 20, and in the former month it endures from 11 P.M. to 9 A.M., while the next highest temperature line, that for 40°, covers in January the time from half-past six in the evening to ten next morning.

It is evident that an attentive study of the diagram will show many interesting facts as to the climate of London. Thus, the coldest time in summer is from about three to five in the morning, while in winter there is not much change between four in the afternoon and eleven the next morning.

In winter the entire range does not exceed 5°, while in summer it amounts to about 16°, and at the equinoxes, in spring and autumn, to 10° or 12°.

It will also be noticed how much colder the spring equinox is than the autumnal, for on the first of April the temperature ranges from 40° to 50°, while on the 1st of October the range is from 50° to 61°.

E

CHAPTER IV.

RADIATION.

WE have hitherto been speaking of the temperature of the air, but have not alluded to the source whence that temperature is derived, or to any means of measuring such an agency. The heat which affects our atmosphere must be derived from one of three sources, the earth itself, the stars, or the sun.

Although the earth possesses internal heat, as is shown by the fact that its temperature increases with the depth below the surface, at the rate of about one degree for every fifty feet, we are justified in disregarding this internal heat as an agency which may exert influence on our climate, for the simple reason that the earth is not growing appreciably colder. If the earth gave out heat to the atmosphere around it to any measurable extent, its own temperature must necessarily fall, and if that fell the mass of the earth would grow colder and would contract. Now the effect of any contraction would be to make the earth revolve more quickly on its axis, or to make the day shorter. Astronomers, however, tell us that the day has certainly not become shorter within historic times; in fact, that the tendency is for it to grow surely but very slowly longer, owing to the retarding action of the tides. The earth's temperature, therefore, cannot have been appreciably reduced within the last three thousand years, and we may assume that the earth, as an independent source of heat, may be practically disregarded.

The relation between the earth, considered as an independent source of heat, and the sun, is thus stated by Dr. Haughton:[1] 'The heat received from the interior of the earth, at present, is sufficient to melt a layer of ice one quarter of an inch in thickness all over the surface of the globe; while that received from the sun would melt a layer forty-six feet in thickness; being thus 2,208 times greater than the heat derived from the interior.'

It need scarcely be said that the amount of benefit in the way of heat we receive from the stars is infinitesimal, for the heat which reaches us even from the moon, which is much nearer than the stars, is so small in amount as to be hardly measurable even by the most delicate apparatus.

We see, therefore, that we must regard the sun's rays as the sole source of the heat which renders the earth habitable, and radiation from the earth into space as the sole way in which the heat is again lost.

Solar Radiation.—Our first problem, therefore, is to measure the amount of solar heat when it has arrived within our atmosphere. For this purpose various plans have been proposed, but none of them can be described as thoroughly satisfactory. It is not sufficient to expose a thermometer to the rays of the sun, for, if the bulb be bright, much of the heat which strikes it will be radiated from it again.

It would go beyond the scope of this work to de-

[1] *Six Lectures on Physical Geography*, p. 77.

scribe particularly Pouillet's Pyrheliometer, one of the first instruments devised to measure solar radiation. A more recent instrument for the same purpose is Herschel's Actinometer, which consists of a glass tube furnished with a bulb, like a thermometer, but filled with a highly-coloured blue liquid, an ammoniacal solution of copper, of which the expansion affords a measure of the sun's action, and provided with an arbitrary scale. The instrument is placed inside a blackened box, covered by a lid, removable when an observation is to be taken. The front of the box is glazed, to protect the bulb from being cooled by convection, as it would be if the air had free access to the bulb. The instrument is taken out and exposed in the sunshine for a minute, and then in the shade for an equal interval. These operations are repeated, and the mean of all the indications is taken as the measure of the desired effect. An arrangement is provided by which the column is brought to the zero point before each observation by means of a screw. The instrument is very difficult to manage, for it is so sensitive that a very slight elevation of temperature fills the entire tube with liquid. It is now rarely used.

One of the most recent forms of actinometers, which has been found to yield good results, is that invented by Professor Balfour Stewart, F.R.S., ('Nature,' vol. xiii. p. 118). The arrangement adopted is to enclose a large-bulb thermometer in a cubical cast-iron chamber of such massive material that its temperature will remain constant for some time. The sun's rays are admitted to the thermometer bulb through a hole in the chamber wall. The mode of making the observation with this instrument is generally similar to that employed with Herschel's instrument.

The use of a box, blackened inside and out and provided with a glazed lid, to increase the effect of the sun's heat on an inclosed thermometer is a very old device. The reason of its efficacy is that while the glass is *diathermanous* to the rays of heat coming from the sun, it is nearly *athermanous* [1] to rays coming from a cooler body, such as the contained thermometer, and it therefore does not allow the heat which enters the box to escape with the same facility as it passed in.

A familiar illustration of this property of glass is afforded by the use of glass fire-screens, which allow a view of the fire while stopping its heat, whereas, as everyone knows, the glass of a window stops comparatively little of the sun's heat. The reason is that the temperature of the fire is infinitely lower than that of the sun's surface, and accordingly the so-called 'dark' heat rays from the fire cannot pass through the glass, while those emanating from the sun are freely transmitted.

In such a box as has been described the temperature has been known to rise above the boiling-point of water, but it is obvious that such a method of observation can lay but little claim to scientific accuracy.

The instrument generally adopted in this country for measuring the intensity of solar radiation is termed the Black Bulb Thermometer *in vacuo*, and its construction was first suggested by Sir John Herschel. This instrument in sunshine will show a temperature far above that of an ordinary thermometer placed beside it.

It consists of a sensitive maximum thermometer

---

[1] *Diathermanous* bodies are those which allow heat to pass through them, *Athermanous* those which do not.

having the bulb and about an inch of the stem coated with dull lamp-black.[1] The whole is then inclosed in a glass tube, of which one end is blown out into a bulb of about $2\frac{1}{4}$ ins. diameter, in the centre of which the bulb of the thermometer is fixed. The glass jacket, so constructed, is then exhausted of air by a good air-pump, and permanently closed.

It is evident that as the action of this instrument depends in some measure on the completeness of the vacuum, it is not sufficient simply to test the thermometer which is to be inclosed in the envelope. It is necessary to have some independent method of gauging

FIG. 9.

Black Bulb Thermometer *in vacuo*.

the amount of rarefaction which has been attained. This may be done in various ways, some makers introducing a mercurial pressure gauge into the vacuum chamber, while others test the state of the vacuum by the passage of the electric light, by soldering platinum wires into the tube, as shown in the figure.

The instrument is then freely exposed to the sun and air by fixing it horizontally above the ground at the same height as that at which the shade thermometers are placed. This is usually four feet. It must be at a distance from walls or trees, and from any

[1] The object of coating part of the stem is to prevent the temperature of the blackened bulb being lowered by contact with the cooler glass of the unblackened stem.

objects which may obstruct the full rays of the sun, or reflect heat rays to the instrument. The bulb is usually directed to the south-east in these islands.

The reading of such an instrument depends on the degree of temperature at which equilibrium is established between the heat produced by the direct rays of the sun, and the cooling produced by the radiation of heat from the bulb to the external jacket. This latter has a temperature depending on, and differing but slightly from, that of the air surrounding the instrument.

The ordinary rule for using the instrument is to observe the maximum temperature which it registers, to subtract from this the maximum temperature given by a thermometer in the shade, and to set down the difference as the greatest amount of solar radiation indicated during the day.

It has also been suggested to place alongside of the black bulb *in vacuo* a thermometer similarly mounted, and of precisely similar construction, except that its bulb is left bright, and to register the difference in reading between the black bulb and the bright bulb as the amount of solar radiation—the black bulb, of course, reading much the higher on account of its lampblack coating.

The great objection to the use of this black bulb thermometer is that serious difficulty is experienced in finding two thermometers which will be thoroughly consistent in their indications. This seems to be due to differences in the coating of the thermometer bulb, in the glass of the outer jacket, or, finally, in the perfection of the vacuum.

At best, however, this instrument only gives the

highest temperature measured in the sun's rays, without any reference to the duration of sunshine during the day, which latter element, however, may be determined as described, p. 58.

It is hard to see what is the object of continuing to enter the amount of solar radiation in the way just described, for the epochs of the highest temperature reached in sunshine and of the highest shade temperature may be widely different, so that the figure set down does not necessarily represent a physical condition existing at any instant of the day.

The absence of any simple method of measuring the total effect of the sun's heat during the day is a serious defect in meteorological observations, and the invention of an instrument for the purpose, suitable for use at ordinary stations, is highly desirable.

We frequently see in this country temperatures of 150° and upwards recorded by the black bulb thermometer, while in very dry climates, such as that of Leh in Ladakh, to the north of Cashmere, at an elevation of 11,000 feet, the readings have gone up to 214° and even higher. In fact, the amount of solar radiation is influenced to a very large extent by the local hygrometric condition of the atmosphere. Professor Tyndall explains this fact on the principle that aqueous vapour is nearly impervious to radiant heat, as will be explained in Chapter VI.

It is a well-known phenomenon that, at considerable elevations above the sea-level, where the denser and damper portion of the atmosphere is beneath us, the direct effect of solar heat is quite disproportionate to the temperature of the air. In such localities, as, for instance, at Davos, in Switzerland, at the level of

5,000 feet, you can sit in the sun comfortably without a great-coat; while in the shade close by, the temperature is several degrees below the freezing-point.

In high latitudes the same paradox is observed, where the extreme dryness of the atmosphere is due to the intense cold. The observation is as old as the time of Scoresby, that on board a whaler you may see the pitch bubbling out of the seams of the ship where the sun shines on them, while ice is forming on the side of the ship which is in shade.

In the passage of the sun's rays through the atmosphere the loss of the heat which they contain by absorption is more than 20 per cent. on a vertical, and almost all on a nearly horizontal, beam. The sun is therefore hotter at noon than in the morning or evening. The difference is brought about by the greater distance the ray has to travel through the dense, damp, lower strata of the air at sunrise or sunset, than at noon, when the sun has its greatest altitude.

On the whole, we may assume that about one quarter of the heat which reaches the outer limit of the atmosphere is lost before it arrives at the sea-level. The most recent proof of this statement is found in the experiments of M. Violle, of Grenoble, who gives the following table for the amount of heat received from the sun at different elevations above the sea, in Europe, showing that a quarter of the heat disappears in the lowest 15,000 feet:—

Locality	Altitude in Feet	Heat received in Percentages
Limit of the atmosphere	?	100
Summit of Mont Blanc	15·781	94
Grands Mulets	10·007	89
Glacier des Bossons	4·000	79
Grenoble	700	71
Level of Paris	200	68

When we come to deal with the distribution
perature on the globe we shall return to this su

We have said that a measurement of the (
of sunshine is very desirable, and a very simple
of effecting this was devised by Mr. John F. C;
F.G.S., and is now in use at about thirty statio1
British Isles. It consists in the exposure to t
rays of a sphere of glass, which acts as a len

Sunshine Recorder.

image of the sun formed by this sphere is recei\
a strip of mill-board stretched in a frame at th
focal distance. When the sun shines a hole is
the mill-board, and, as soon as the sun sets or i:
by a cloud, the record ceases. It is obvious t]
a method gives no measure of the intensity
radiation, but merely records the length of
which the sun shone with sufficient intensity
a sheet of mill-board. The apparatus is, }
reasonably cheap and very easily managed. T

employed by the Meteorological Office, being that devised by Professor G. G. Stokes, F.R.S., is shown in the figure.

Terrestrial Radiation.—We now come to the measurement of the action converse to solar radiation, the escape of heat from the earth to space. This action is constantly going on, as may be shown by the following experiment :—Place a thermometer with its bulb in the focus of a concave mirror, which is deep enough to cut off all view of the earth from it. If now the axis of the mirror be directed to the blue sky, the thermometer, if screened from the sun's rays, will fall several degrees, while if the axis be directed to a cloud or to any terrestrial object—even though that be, as Herschel proved, the summit of a snow-clad mountain—the temperature will rise. In the former case, so to speak, no heat is transmitted to the thermometer from the clear sky, while in the latter any, even the slightest, obstacle is sufficient to reflect back the heat escaping from the earth. Thus, for instance, von Buch says of Teneriffe that at Santa Cruz and Funchal the temperature does not fall much at night time, as these places are surrounded by hills which check the free radiation, while on the plain of Laguna, close by, the night temperature in winter often falls below the freezing-point.

The instrument employed for the purpose of measuring terrestrial radiation is a simple minimum thermometer, of which the stem is generally inclosed in a glass tube, for some slight security against accidental fracture.

By some authorities it is recommended to use a black bulb minimum thermometer for this observation, but the precise utility of blackening the bulb is not apparent.

It is, however, not at all an easy matter to ensure that the observations of terrestrial radiation shall be strictly comparable at all stations and at all seasons of the year. The rule given is to place the thermometer on greensward with its bulb just on the level of the tops of the grass. In many places, even in this country, it is not easy to ensure that grass shall be always available in winter, and that in summer it shall not be allowed to grow too high.

When speaking of temperature I have said that the effect of solar radiation varies with the nature of the surface on which the rays fall, being least on ice or water and greatest on a sandy soil, and instances of the extraordinary temperatures attained under the last-named conditions have been cited. In proportion as the ground is covered with, or permeated by, water, is the effect of solar radiation, in warming it, reduced, for a large proportion of the heat which reaches it is employed in evaporating some of the water, and is, therefore, rendered latent.

The facility with which heat is radiated from the earth into space is affected to a great extent by the nature of the covering with which the earth is provided. Grass and herbage are better radiators than earth or gravel, and, accordingly, the temperature at night time will fall far lower over a meadow than over the road passing beside it. The action of trees is, of course, much more energetic than that of grass, in consequence of the great extent of radiating surface afforded by the leaves.

The indications of a thermometer placed just over grass will, therefore, as a general rule, range, during the night time, when radiation from the earth is most active, several degrees below those of a similar instrument placed in a screen above the grass, and protected from the action of radiation. As a general rule, the grass minimum thermometer reads several degrees below the ordinary minimum at the height of 4 feet above the ground. In some instances, however, as, *e.g.*, during very wet fogs, it has been found that the temperature on the grass has been higher than that in the screen; but this phenomenon is very rare.

In order to secure a surface of uniform radiating power in all climates and at all seasons, it has been suggested to lay the thermometers on a piece of cloth and dispense with the use of grass altogether. The question of nocturnal radiation is, however, very complicated, and requires much further investigation; we shall see its importance when treating of dew and hoar-frost in Chapter VII. A careful and elaborate discussion of the subject will be found in a paper by Mr. Jas. Glaisher, 'Phil. Trans.,' 1847.[1]

A very practical use of nocturnal radiation has been made from time immemorial in India in the preparation of ice, and on such a scale that about 10 tons of ice can be procured in a single night from twenty beds of the dimensions about to be given, when the temperature of the air is 15° or 20° above the freezing-point. An account of the practice followed will be found in a paper by Mr. T. A. Wise, given in 'Nature,' vol. v., p. 189. The locality referred to is the immediate neighbourhood of Calcutta. A rectangular piece of ground is marked out, lying east and west, and

---

[1] 'On the Amount of the Radiation of Heat at Night from the Earth, and from various bodies placed on or near the surface of the Earth.' By Jas. Glaisher, *Phil. Trans.* 1847, p. 119.

measuring 120 by 20 feet. This is excavated to the depth of two feet, and filled with rice straw, rather loosely laid, to within six inches of the surface of the ground. The ice is formed in shallow dishes of porous earthenware, and the amount of water placed in each is regulated by the amount of ice expected.

In the cold weather, when the temperature of the air at the ice-fields is under 50°, ice is formed in the dishes. The freezing is most active with N.N.W. airs, as these are the driest; it ceases entirely with southerly or easterly airs, even though their temperature may be lower than that of the N.N.W. wind.

No ice is formed if the wind is sufficiently strong to be called a breeze, for the air is not left long enough in contact with the bed for its temperature to fall sufficiently.

The rice straw, being kept loose and perfectly dry, cuts off the access of heat from the surface of the ground below it, and, when the sun goes down, the straw being a powerful radiator, the temperature of the air in contact with the dishes is reduced some 20° below that prevailing two or three feet above them. The rapid evaporation of the water into the dry air above creates also an active demand for heat to be rendered latent in the formation of steam, and the result of all these agencies is the formation of ice, under favourable circumstances, on the extensive scale above mentioned.

CHAPTER V.

PRESSURE.

THE pressure of the atmosphere comes next in importance to its temperature as the cause of many phenomena of great consequence to human life.

This property of the atmosphere was first discovered by Torricelli, a pupil of Galileo's, who also devised the means of measuring it. Galileo had noticed that water would not rise in a pump over a certain height, eighteen cubits (*diciotto braccia*), above the level of the well; and he endeavoured to explain this fact by comparing the column of water in the pump-bore to a cylindrical rod held by its upper end, which by its own weight lengthens itself and at last breaks.

Torricelli, in 1643, however, devised the following experiment which decided the question. He took a tube about 3 feet long, closed at one end, and filled it with mercury. He then inverted this, and plunged its lower end into a basin filled half with mercury and half with water. When the open end of the tube was immersed in the mercury, the level of the mercury in the tube sank until the length of the column, measured from the surface of the mercury in the basin, was about 30 inches, leaving the space in the tube above the column empty. When, however, the open end of the tube was raised above the level of the mercury, but so as to be still under the surface of the water, the contained mercury immediately fell out of the tube, and its place was taken by the water, which rushed in with great violence and entirely filled the tube. This experiment proved that the height of the column of liquid which would stand in any tube depended on the specific gravity of the liquid of which the column was composed.

The reason why a fluid stands at all in such a tube is that the air presses downwards with a uniform force on all parts of a free fluid surface. If, then, we relieve a portion of that surface from the pressure by inverting an exhausted tube over it, the fluid will be forced up in a column over the space whence pressure has been removed, *i.e.* into the tube, until the weight of that column exercises a pressure equal to that of the air outside.

This pressure is, speaking generally, about 14·7 lbs. on each square inch, and as the specific gravity of mercury is 13·59, and the weight of a cubic inch of water is 252·5 grains, an easy calculation will show that about 34 feet of water, or 30 inches of mercury, will counterbalance the pressure of the atmosphere.

Use is made of the pressure of the atmosphere in the construction of the pump. The air is drawn out of the bore by the sucker, and the water rushes up after it to fill the vacuum produced. In the case of the barometer, we do not draw the air out of the tube, but we fill the tube with mercury and invert it in a basin of the same liquid, when, as above explained, the length of the column supported in the tube will be about 30 inches.

I have said that the space left in the tube above the

top of the column in Torricelli's experiment, when the mercury is used, is empty, and this space has been called the Torricellian vacuum. If the tube be quite dry, this space will contain nothing but a little of the vapour of mercury, and up to very recent times it was considered the most perfect vacuum which could be obtained.

The actual proof that Torricelli's was the true explanation of the principle of the barometer was given by Pascal soon after the original discovery. It struck him that if the height of the column was due to the pressure of the air at the surface of the ground on a plain, this height ought to be less at the top of a mountain than at its foot, owing to the diminished thickness of the stratum of air above the instrument at the summit. His brother-in-law, Perrier, in 1648, ascended the Puy de Dôme in Auvergne, which rises to a height of about 3,500 feet above the surrounding country, and found that this idea was perfectly correct. The barometer read nearly four inches lower on the top of the mountain than in the garden at Clermont. Soon after this experiment was carried out, Boyle proposed for the instrument the name of 'Barometer,' which has since been universally adopted.

The simplest form of the instrument consists of a tube of large bore standing vertically in a vessel of mercury. The height of the column in the tube above the level of the mercury in the cistern is measured by means of a graduated scale placed beside it. If great accuracy is required, this scale is read by means of a telescope at a distance, fixed in a framework called a cathetometer (a measurer of perpendicular height), which ensures its perfect horizontality at all levels. This is the

F

form of barometer used as the standard at Kew Observatory.

The glass tube must be fixed in a frame to protect it from injury. In ordinary barometers the scale is made a part of the instrument, being either marked on the frame itself or on a rod attached to it. It is sometimes engraved on the glass tube.

The divisions of the scale need not be carried out along the whole length of the tube, for the readings at sea-level will range between 26 and 32 inches. If, however, the instrument be wanted for stations at high levels, the engraved part of the scale must extend below 26 inches. Mr. Glaisher, in the famous balloon ascent mentioned (p. 12), actually recorded a reading of $9\frac{3}{4}$ inches, at the height of above 29,000 feet, which had been reached when he lost consciousness. The balloon mounted higher, and Mr. Coxwell believed that he himself noticed a reading of 7 inches before the descent commenced. This would give a height of 37,000 feet.

In all considerations of the question of its expansion or contraction by heat it is held that the scale consists of the material of which the frame is made. In almost all instruments used for scientific purposes this material is brass, and, as the expansive co-efficient of brass is well known, it is possible to calculate the effect of heat in altering the length of such a scale—an element which has to be taken into consideration in reducing barometer readings to their equivalents at a fixed temperature, as we shall see presently.

If the barometer be in a wooden frame, unless the scale be on a rod, independent of the frame, it must be considered to consist of wood, though the figures may be engraved on ivory, glass, or porcelain. A wooden scale is seriously altered in length and shape by changes in the dryness of the air, and wood is not a homogeneous material. Of such alterations no account can be kept, and therefore barometers in wooden frames are of no use for scientific observations.

This statement, of course, does not apply to barometers in brass frames attached to wooden boards, or placed in outer glazed cases of wood.

Barometer Vernier.—In order to facilitate the taking of accurate readings of the height of the barometer, a small movable scale (fig. 11, p. 69), called a ' Vernier,' from the name of its inventor, is attached to the instrument.

The general principle of this device is that a given length, containing n divisions of the fixed scale, is divided into $n \pm 1$ divisions on the vernier. In most barometers this number n is 24, and the distance occupied by 24 spaces on the fixed scale is divided into 25 spaces on the vernier. Each space on the fixed scale is 0·05 in. and each such space is larger than a space on the vernier by the twenty-fifth part of 0·05 in., which is 0·002 in., so that the vernier shows differences of two thousandths of an inch.

The usual graduations of the scale and vernier for English barometers are as follows:—

The scale is divided into inches, tenths, and five hundredths of inches. The inches and half-inches are marked by bolder lines after them.

Every long line on the barometer scale corresponds to a tenth, 0·100, of an inch.

Every short line on the barometer scale corresponds to five hundredths, 0·050, of an inch.

Every long line on the vernier scale corresponds to one hundredth, 0·010, of an inch.

Every short line on the vernier scale corresponds to two thousandths, 0·002, of an inch.

Reading the Barometer.—The vernier is moved by a rack and pinion. The milled head of the pinion should be turned so as to bring the *lower* edges of the vernier exactly on a level with the top of the mercurial column, which is usually more or less convex. When set properly, the front edge of the vernier, the top of the mercury, and the back edge of the vernier should be in the line of sight, which will thus just touch the uppermost point of the column. Great care should be taken to acquire the habit of reading with the eye exactly on a level with the top of the mercury. If the vernier be brought down too far its edges will correspond with a chord of the curve formed by the surface of the mercury, instead of being a tangent to that curve; and the reading will be too low.

A piece of white paper placed behind the tube, so as to reflect the light, will be found useful in setting the vernier; and at night a small bull's-eye lantern, or a candle, held at the side of the instrument so as to throw the light on the paper, will enable the observer to take a correct reading.

When an observation is being made, the barometer should hang quite freely; it should not be held or touched, for fear of its being moved out of the perpendicular. The least inclination will cause the column to rise in the tube.

Just before setting the vernier, one or two taps should be given to the instrument by the hand, of sufficient force to cause the top of the mercurial column to

PRESSURE. 69

be agitated. This operation overcomes the tendency of the mercury to adhere to the glass, and allows the force of capillarity to exert its normal action. If ever the difference of height thus caused is of any importance it is when the pressure of the air is rapidly changing, and the barometer has been kept perfectly undisturbed for a time. A beginner should watch the effect of tapping by a practised observer, in order to learn the amount of force he may safely apply without risk of damage to the instrument.

It may be useful to describe the actual mode of reading off. In the following diagrams, AB represents part of the scale, and CD the vernier, the lower edge of which, D, denotes the position of the top of the mercurial column. The scale is easily understood:—B corresponds to 29·000 ins.; the first line above B, a short line, is 29·050 ins.; the second line, a long line, is 29·100 ins. and so on. The scale line, next below D, is first noted by the observer, and he then looks to see which line of the vernier is in one and the same straight line with a line of the scale.

FIG. 11A.   FIG. 11B.

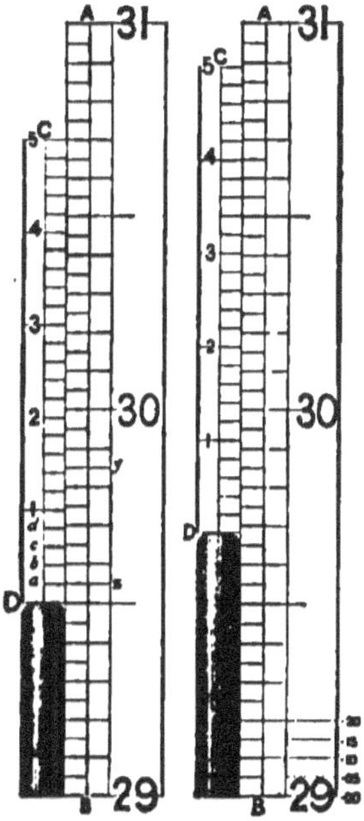

Setting of Vernier.

In fig. 11A, the lower edge of the vernier D is supposed to coincide with the scale line 29·5. The reading is therefore 29·500 ins. If we examine the figure carefully we see that the vernier line a is ·002 ins. below the next line z of the scale. If, therefore, the vernier were moved so as to bring a in a line with z, the edge D would read 29·502 ins. In like manner it will be seen

that b is ·004 inch
,, c ,, ·006 ,,
,, d ,, ·008 ,,
} away from the line on the scale next above it,

and that 1 on the vernier is ·010 in. below y on the scale. If, therefore, 1 were removed into a line with y, the edge D would read 29·510 ins. Thus the numbers, 1, 2, 3, 4, 5, on the vernier indicate hundredths, and the intermediate lines on the vernier indicate the even thousandths of an inch.

If we now look to fig. 11B (p. 69), we find that the scale-line next below D is 29·650 ins., and that the third line above the figure 3 on the vernier coincides with a line on the scale. The number 3 indicates ·030 in., and the third subdivision ·006 in., and thus we get

Reading on scale	29·650
Reading on vernier	$\left\{\begin{array}{l}·030\\·006\end{array}\right.$
Correct reading of barometer . .	29·686 ins.

Sometimes two consecutive lines on the vernier will appear to coincide with two lines on the scale; in this case the intermediate thousandth of an inch should be set down as the reading. Thus, if the reading appears to be 29·684 ins. or 29·686 ins., the mean, 29·685 ins., should be adopted as the true reading.

Attached Thermometer.—Every mercurial barometer should have a thermometer attached to its case, the bulb of which should be turned inwards, so as to be as near as possible to the barometer tube. The thermometer may also be placed in an open glass tube of the same section as the barometer tube, attached to the case of the barometer and filled with mercury. The attached thermometer is an indispensable adjunct to the instrument, as no reading of the barometer is complete without an observation of its temperature at the same time; and this observation should be taken before the barometer itself is read, for the proximity of the observer's person may cause the attached thermometer to rise so that it will not indicate the true temperature of the mercurial column.

In handling barometers it should be remembered that they are easily damaged by rough usage, and it is not always easy to have them properly repaired. Observations from an instrument imperfectly repaired or unverified after repair are useless for scientific purposes. For travelling, the mercury should be caused to rise to the top of the tube so as to fill it entirely (the mode of effecting this differs with different classes of instruments). The barometer should be carried with the cistern end upwards; and if packed in a box, care should be taken that the lid is screwed, not nailed on, for the hammering may easily break the tube.

The barometer may be placed in any convenient room where it is not exposed to accidental injury from blows of passers-by. It must be hung in a perfectly vertical position and kept vertical by clamping screws, It should be at such a height that the top of the column can be easily read off by a person standing by. It should be in a good light for observing, but out of the reach of sunshine and (as much as possible) out of the action of the direct heat of a fire. It should not be exposed to sudden changes of temperature which might affect the readings of the attached thermometer, and prevent it from indicating the temperature of the mercurial column.

FIG. 12.

I shall now proceed to describe some of the principal forms of barometer.

Standard Barometers.—We have already spoken of barometers in which the scale is marked off on the case. In the most perfect form of these instruments the starting-point of the scale is formed by a pin of ivory, and this, when a reading is being taken, must be brought into exact contact with the surface of the mercury in the cistern, as the lower vessel is called. It is obvious that if this adjustment be properly made, and the scale be properly laid off from the ivory point, the only possible sources of error in the resulting reading will be those arising from the capillary depression of the mercury in the tube, and those produced by the temperature of the mercurial column. The mode of correction for these two errors will be explained subsequently.

The adjustment of which we have spoken is usually effected by screwing up the leather bottom of the cistern until the level of the contained mercury is raised so that the actual ivory point and its image, as reflected in the bright surface of the metal below, appear exactly to touch each other.

Fortin's Barometer.

The instrument thus described is Fortin's Barometer (fig. 12), and its use involves two operations; one, the setting of the ivory point, the other the reading of the top of the column. The adjustment of the point necessitates the admission of light to the surface of the cistern by means of a glazed aperture cut in the cistern wall.

If this light were not admitted it would still be possible to obtain an accurate reading, for if we knew the so-called 'neutral point' of the scale, and also the ratio of the sections of the tube and the cistern, we could, by applying a 'capacity correction' calculate the true height of the mercury for any observed reading, inasmuch as when the barometer falls any mercury leaving the tube must enter the cistern, and *vice versâ* when the barometer rises. However, we need not pursue this subject further, for practically barometers requiring the capacity correction are not now made.

There are three ways of escaping the necessity for a capacity correction, one being the adoption of Fortin's pattern for the instrument, which, however, as we have seen, involves the taking of a double reading. Another avoids this difficulty, but renders the reading absolutely dependent on the correctness of an artificial scale, without affording the observer any means of directly testing the accuracy of the instrument he is employing.

This pattern of instrument is that invented by Mr. P. Adie in 1854, for use at sea, and commonly known as the Kew Barometer. In it the extreme length of the scale is marked on the instrument, but instead of laying off true measurements on it the inches are shortened from the upper part downwards in proportion to the relative sizes of the diameters of the tube and cistern; the highest point on the scale (say 32 inches) being marked off correctly from a definite point on the cistern side. As in the ordinary Kew barometers these diameters measure about 0·25 in. and 1·25 in. respectively; the inches of the scale are shortened in the proportion of 0·04 inch to one inch.

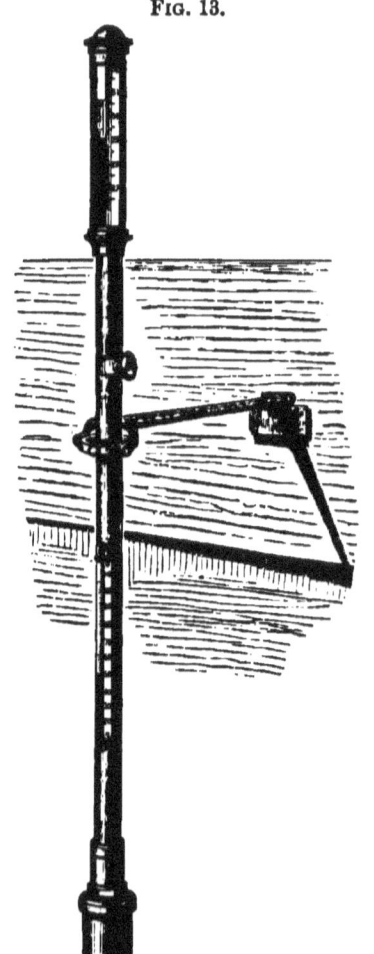

Fig. 13.

Marine Barometer. Kew Pattern.

The Kew barometer (fig. 13) therefore only requires one reading instead of two in order to ascertain the pressure; but, as has been said above, its correctness depends on the accuracy with which the scale has been laid off. This can only be ascertained by actual experimental alteration of the pressure by means of an air-pump, and comparison with a standard barometer under similar circumstances of pressure. Its error at each half-inch can then be noted.

These barometers are coming more and more into use every day. In their original form they were intended for use at sea. In Marine barometers, however, the tube is contracted for the greater portion of its length, in order to check the irregular oscillations of the mercury, due to the motion of the ship, a defect technically known as 'pumping.' This contraction of the tube renders the instrument somewhat sluggish, but not sufficiently so to form a very serious objection to its use at land stations.

In fact, from one point of view, this contraction is a great recommendation for these instruments, as it renders the instrument much less liable to damage in transit, owing to there being comparatively little mercury in it, and to the freedom of motion of that mercury being much restricted.

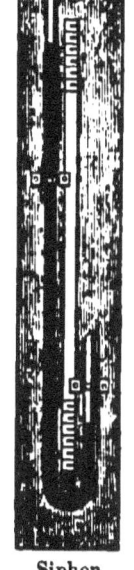

Fig. 14.

Siphon Barometer.

The third method of getting over the necessity for a capacity correction is very extensively employed abroad, and it presents some advantages, especially for travelling. It consists in dispensing with the cistern altogether, and using a U-shaped tube. This is the Siphon Barometer (fig. 14), in which the long leg is closed and the short leg open. In this instrument as the mercury rises in the long leg it must sink in the short, and *vice versâ*. The motion in each leg is exactly one half of what takes place in a Fortin's barometer. There are always two readings to be taken of the respective heights of the mercury in the two legs of such an instrument. There is, however, no need to apply a capillarity correction, p. 84, to either reading, as the two legs are equal in section and the action of capillarity is the same in both. In the opinion of some meteorologists the siphon is the best form of barometer, for in its use the two observations taken are both of the same phenomenon, the height of the mercury in a tube; whereas, in the

Fortin, one is the adjustment of an ivory point to touch the surface of mercury in a vessel, the other an observation of the mercurial level in a tube; two totally distinct phenomena.

Another advantage attaching to the use of siphon barometers is that if an observer has a tendency to read too high or too low, the error will affect both readings, which will therefore compensate each other. This form of instrument, from its lightness and portability, is the only form of mercurial barometer which can be used for mountain ascents; but the inconvenience entailed in ordinary use on the observer by the double reading has effectually displaced it, at least in these islands, from public favour, as compared with the Fortin and Kew barometers.

Siphon barometers are prepared for removal by sloping them. This causes the mercury in the long leg to rise to the top, and that in the short leg passes beyond the bend in the tube, so that no air can possibly get up as long as the instrument is carried with the bend uppermost.

The ordinary wheel barometers, so commonly used, are constructed in the following way. A float is placed on the mercury in the open leg of a siphon barometer, and this rises or falls as the column in the tube falls or rises. The motion is transferred, by means of a string and block, to a hand like that of a clock, and the scale of the instrument can be thus magnified to any extent that may be wished.

The motion of a float may be also employed to give an automatic record of the height of the mercury, by attaching to the string a pencil, which makes a mark on ruled paper, moved by clockwork. The mark may be made either at regular intervals or continuously.

Such an automatically registering instrument is termed a *barograph*. The principle of registration generally adopted in this country for the better class of barographs is photographic, not mechanical.

There is this objection to all photographic barographs, that the record is invisible until it is developed, so that it is impossible to see what changes in pressure are in progress at any time when a phenomenon of interest, such as a heavy squall, may be experienced.

In the barograph in use at Greenwich, devised by the late Mr. Charles Brooke, F.R.S., the barometer is of the siphon pattern, and an arrangement is provided by which a continuous record of the varying position of the float is obtained by photography on a sheet of paper moved by clockwork.

In the barograph employed at the observatories in connection with the Meteorological Office, which was devised by the late Sir F. Ronalds, the barometer is of the ordinary pattern, and the light is admitted through the Torricellian vacuum, so that the actual height of the mercury itself is photographed, without the intervention of any mechanical contrivance.

As a fact all instruments which are provided with a mechanical contrivance for registration must necessarily be more or less sluggish in their action as compared with the simple instrument, for the mercury has not only to move in accordance with the changes of pressure, but to do the mechanical work of moving the float.

The most satisfactory mechanical siphon barograph as yet invented is that of Redier, in Paris, described in the 'Quarterly Journal of the Meteorological Society' (vol. ii. p. 412). In this instrument, the inventor says, 'the apparatus is so arranged that the work is done by a powerful clock movement, and the barometer has only to direct the action of the clockwork.'

The most perfect form of barograph, yielding, by means of a pencil, a continuous record on which, therefore, the changes of pressure can be observed as they take place, is that of King, which has long been employed at the Bidston Observatory, at Liverpool. The principle of this instrument entirely differs from that of any barometers hitherto mentioned. It is fully described in Mr. Hartnup's 'Report to the Mersey Docks Board for 1865.'

It is evident that the column of mercury in a barometer tube has a certain weight, and this weight is increased when the column rises, and decreased when it falls. If the tube be suspended from the arm of a balance with its open end immersed in a cistern of mercury, and exactly counterpoised, then if its weight increases, the tube will sink until the height of the column left above the mercury in the cistern is so reduced that its weight is exactly equal to that of the given counterpoise; and, *vice versâ*, if its weight decreases, the tube will rise. This motion of the tube furnishes abundant mechanical force to move a pencil and register the changes of level produced by the alterations in pressure. There is no necessity for any temperature correction in this barograph, for the weight of mercury required to counterbalance the pressure of the atmosphere does not depend on the temperature of the column. It is evident that the scale of such an instrument may be increased almost at will, so that the minutest oscillations of pressure may be recorded.

Before leaving the subject of the measurement of pressure we must not omit to mention the aneroid barometer, which has come into very extensive use, owing to its convenient size and portability. In this instrument atmospherical pressure is measured by its effect in altering the shape of a small, hermetically sealed, metallic box, from which almost all the air has been withdrawn. When the atmospherical pressure rises above the amount which was recorded when the instrument was made, the top is forced inwards; and,

Fig. 15.

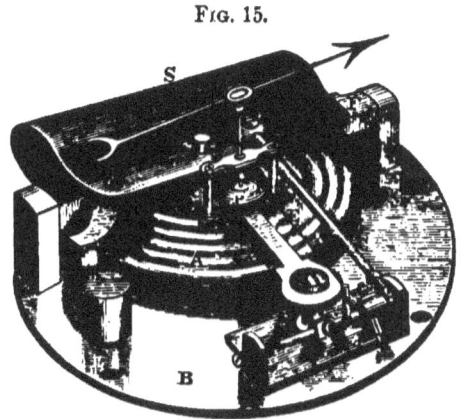

Construction of the Aneroid Barometer.

vice versâ, when pressure sinks below that amount the top is pushed outwards by a spring.

The instrument (fig. 15) consists of the vacuum chamber, A, composed of two discs of corrugated German silver, soldered together and forming a box. This is attached by a pin to the base plate of the instrument, B, and by another to a strong spring, S, which is borne by the frame, F. The spring acts in opposition to the motion of the vacuum chamber. The lever, C, attached to the spring is composed of iron and brass, and is intended to compensate for the effects of temperature. It is connected, by means of the bent lever at D, to a chain, E, which is wound round the arbor, O. The spiral spring, P, keeps the chain wound round the barrel free from slackness while the pressure is decreasing, and the hand moves to the left over the dial. When the pressure is increasing the lever pulls upon the chain, and the hand moves to the right over the dial. The hand is attached to the arbor and moved by its rotation.

The principle of Bourdon's metallic barometer is somewhat similar to that just described.

It is evident that these instruments must be graduated experimentally, as they cannot measure pressure absolutely, but only afford indications relatively to a mercurial barometer. Almost all aneroids are compensated for temperature, and so the readings of the mercurial barometers which are to be compared with them must be reduced to 32° (p. 85).

Aneroids are very sensitive, but they do not preserve their accuracy, owing either to imperfection of workmanship, to rust or corrosion, or to the alteration of force in the springs used in their construction, and unfortunately the instrument itself may give no indications of being out of order till it is compared with a mercurial barometer. Accordingly, if on a journey or a mountain ascent an aneroid meets with accidental rough usage, all subsequent readings made with it may be quite erroneous, while the instrument shows no signs of having sustained any injury.

Even without any rough usage it will often be found that when a table of corrections has been determined for an aneroid, the instrument may in the course of time have undergone some change, and the values of the corrections will require alteration. Constant re-comparisons of aneroids with standard barometers are therefore desirable.

The consequence is that the aneroid can never be employed for scientific observations requiring great accuracy, though it is very convenient for use as a weather-glass.

The Sympiezometer (from σύν πιέζω μέτρον, a measurer of compression) is an instrument invented by Adie of Edinburgh, and was formerly much used at sea, as being more sensitive than the mercurial barometer. It consists of a short inverted siphon tube, with a bulb blown on the end of the longer limb, while the shorter limb is widened in a cylindrical form, and left open, except when closed with a cork for the purpose of travelling. A small thermometer is also attached, and forms an essential part of the apparatus.

The bulb is filled with air, and the lower part of the tube and the cylinder with glycerine.

If atmospheric pressure remains unchanged, the instrument will act as an air thermometer, the contained air being expanded by heat, and contracting when cooled. If the temperature remains unchanged the air in the bulb will be compressed when pressure increases, and will expand when pressure decreases. The instrument will therefore measure pressure.

Hence we see that in employing the instrument regard must be had to the temperature. In fact, the scale must on each occasion be set for the temperature, before the indication of pressure is read off.

It is found that sympiezometers are very liable to get out of order, and they are now hardly ever used.

Returning to mercurial barometers, we must now mention some of their most common defects. Of these, that which causes the greatest trouble is the presence of air or moisture in the tube of the instrument. It has already been explained that if the tube has been properly filled, the space at the top should contain nothing but a trace of the vapour of mercury, the tension of which, at ordinary temperatures, is so slight as to make no appreciable difference in the height of the column. If, however, the tube has not been perfectly cleaned and dried, before being filled with mercury, any moisture left in it will ascend to the top, and moreover small quantities of air will have a tendency to creep up between the mercury and the inner walls of the tube. The presence of any gaseous substance in the vacuum depresses the mercury, as its elastic force tends to counterbalance a portion of the pressure of the atmosphere outside. If moisture be present, besides the effect of its vapour, it causes the mercury to adhere to the glass, and so makes the instrument sluggish in its action.

If, therefore, a barometer reads too low, or appears to be sluggish, or if the mercury adheres to the glass, the presence of air or moisture may be suspected, and the instrument should be sent to the maker to be set to rights.

It is usual, when a tube has been filled with mercury, to heat it up to the boiling-point of the metal in order to expel the last traces of air and moisture. To ascertain if a tube has been properly 'boiled,' it is generally sufficient to examine it carefully with a lens, when, if no small specks or air-bubbles be detected, it may be assumed that the boiling process has been carried out.

In order to test if there be air or moisture in the tube, it is sometimes recommended to incline it, and allow the mercury to flow suddenly to the top. If then the vacuum be perfect, a sharp click will be audible; whereas if air should be present, the sound will be dull. Great caution is requisite in making this experiment, as, if the barometer be inverted too quickly, the weight of the mercury falling suddenly is very likely to break out the end of the tube.

FIG. 16.

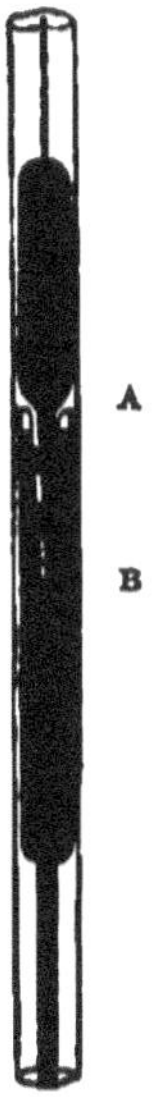

In order to prevent the access of air or moisture to the vacuum, it is the practice to insert a small funnel, or pipette, into the tube somewhere between the scale part of the column and the cistern. This arrangement is called Gay-Lussac's pipette, as he first suggested it, and its effect is to check the ascent of any air, &c., into the vacuum. Any such bubbles will lodge in the shoulder A, as represented by the white space (fig. 16). Experience shows that this arrangement tends to preserve the instrument for a long time. When a barometer is provided with it, air can only get up into the vacuum when the lower end of the pipette is uncovered by mercury, either by so much air having accumulated as to fill the entire space from A to B, or by the end B being left uncovered by mercury when the instrument is laid flat in its box.

The Pipette.

Sometimes a particle of dirt may lodge in the pipette, or in the contracted tube of a marine barometer, and completely check the action of the instrument.

If therefore a barometer remains steady when the mercury ought to exhibit motion, some defects of this character may be suspected. In such a case the instrument should be taken down and inverted, and then laid aside for some time, cistern uppermost. This treatment will probably have the effect of removing the obstruction. If not, the instrument must be sent to the maker to be set to rights.

Barometer Corrections.—We now come to the subject of the correction of barometrical observations. Every observation of the barometer must be *corrected* in order to make it comparable with other observations taken at other times and at other places, and therefore probably under conditions which are more or less different.

Some of these 'corrections' have reference to the individual instrument, others are applicable to all readings of any instrument.

Of the former class there are three: the corrections for A. Index Error; B. Capacity; C. Capillarity.

A. *Correction for Index Error.*—This is discovered when the instrument is verified at an establishment like Kew Observatory. It is, in fact, the error made by the workman who laid off the scale of the instrument.

B. *Correction for Capacity* is only requisite in old-fashioned barometers, as explained at p. 73. Its amount depends on the proportion borne by the sectional area of the tube to that of the cistern. At one point on the scale the reading is correct, and the correction to be applied for capacity is additive when the mercury is above, and subtractive when it is below that point.

C. *Correction for Capillarity.*—The capillary action between glass and mercury has a tendency to depress the column, the amount depending on the size of the tube, and being greater the smaller the tube. It is also greater in an 'unboiled' than in a 'boiled' tube.

All certificates from Kew Observatory, for Kew pattern barometers, give a correction at each half-inch, embracing the three corrections above named.

The corrections which are independent of the special instrument are two in number: D. for Temperature; and E. for Altitude.

D. *Correction for Temperature.*—All bodies are affected in their dimensions by heat, and it is therefore necessary to reduce the mercurial column to some uniform temperature in order to make readings mutually comparable. The temperature universally selected for this purpose is 32° F., and so the correction for temperature is commonly called the *reduction to 32°*. Under ordinary circumstances, in this country, the reduction to 32° makes the readings lower, as the temperature of the room where the instrument is placed is generally above that point. A condensed extract from the best reduction tables will be found in Appendix III.

E. *Correction for Altitude.*—We have seen at p. 65 that the fact that the barometer reads lower on the top of a hill than at its foot has been known for two centuries. It is therefore obvious that readings at different stations must be reduced to their equivalents at the same level. For this the Mean Sea-level (*i.e.* for Great Britain the *mean half-tide level at Liverpool*) is always taken, and so this correction is called *the reduction to sea-level*. In most parts of these islands the height of any station may be found by applying to the Ordnance Survey Office at Southampton, and estimating, if it be not possible actually to measure, the height above or below the nearest point of which the elevation has been accurately determined by the Survey.

The problem of reduction to sea-level is, however, not quite so simple as might be imagined. What is required is to find the weight of the vertical column of air which would extend from the level of the upper station to that of the sea. This weight must depend on the temperature of the air, for when the temperature is high a longer column is required to make the same weight than when it is low, and this temperature is different at the levels intermediate between the two stations, and may be affected by other conditions, as explained in the preceding chapter. It is, therefore, not a simple matter to determine the true average temperature of the column. However, for every-day observations and for moderate heights, such as, for the most part, are met with in these islands, the temperature is assumed to be that of the external air at the station where the barometer is placed.

Tables for the correction of readings for heights up to 1,500 feet, and at various temperatures, are given in the Instructions already mentioned, and an extract from these will be found in Appendix IV. The reduction to sea-level for stations above that elevation must be made by special calculations from the formula there given.

A simple, and approximately accurate, rule for ascertaining the relative elevation of two stations is to multiply by 9 the difference in barometrical readings between them, taken in hundredths of an inch. The result will give the difference in feet between the stations. This rule is sufficient for the ordinary problems as to elevations which occur, *e.g.*, during a walking tour, when the height is roughly measured by means of aneroid readings. This convenient plan was first suggested by Mr. R. Strachan. It depends on the principle that the difference of height corresponding to a difference in barometrical readings of 0·1 inch is approximately 90 feet. For instance, for the change of level from 200 to 290 feet, the difference in barometrical readings (the sea-level reading being 30 inches) is ·101 inch at 40°, and ·098 at 50°.

The barometrical determination of heights in the interior of continents, or at places distant from the sea, is more or less uncertain. In such cases it is hardly possible to take readings at the upper and lower station which shall be exactly simultaneous, and so we are obliged to have recourse to the mean of a series of readings at each station. Now, as we shall presently learn, the mean pressure varies at different parts of the earth's surface; thus, within the limited area of these islands, on the average of the ten years 1866–75, it ranges from 29·808 inches at Nairn, to 29·965 inches at Plymouth. This difference would correspond to a difference in level of more than 150 feet.

If we were to take a station, say in Auvergne in Central France, it would make a great difference in the correction to be applied whether we took the sea-level station on the coast of the Mediterranean, of the Bay of Biscay, or of the English Channel.

On this account, and from the difficulty of ascertaining correctly the distribution of temperature, some meteorologists object *in toto* to employing barometrical readings reduced to sea-level, and prefer to use them simply corrected for temperature, and to take the differences of the individual readings from their normal values as a truer indication of the condition of the atmosphere than a series of reduced observations.

The discussion of barometrical observations is, of course, conducted on the same principles as that of observations of the thermometer, but in the case of the barometer we are perforce driven to take, as the mean for the day, some combination of observations taken at fixed hours, and have no short method of arriving at the desired result by taking the arithmetical mean of the highest and lowest readings during the day, as we possess when investigating temperature.

The curve of diurnal variation of the barometer is fortunately of much less importance than the corresponding curve for the thermometer, and although a regular oscillation does exist, its appearance in these latitudes is rarely perceptible on the trace of a barograph for a single day, unless in very calm weather. In fact, as the changes of pressure due to this cause only amount in these islands on the mean of the year to about 0·02 in., while the change of level produced by the passage of a storm sometimes exceeds an inch in a day, the former motion is generally masked by the latter. In scientific language, the *periodical variations* are concealed by the *non-periodical variations*, the oscillations due to the passage of storms.

The curves of daily variation are, however, a striking feature of the meteorology of the torrid zone, insomuch that any interruption to, or irregularity in, their course is an unfailing sign of a storm of some sort. It is indeed said that in those latitudes you can tell the time of day, within twenty minutes, by an accurate barometrical reading.

The ordinary type of curve exhibits two maxima, which occur at *about* 9 A.M. and 9 P.M., and two minima, which occur at *about* 3 A.M. and 3 P.M., but these epochs are not the same at all stations (fig. 17). At Calcutta, which will serve as a good specimen of a tropical station, the maxima and minima fall respectively an hour later than, for instance, at Greenwich.

As a general rule, the morning maximum and the

FIG. 17.

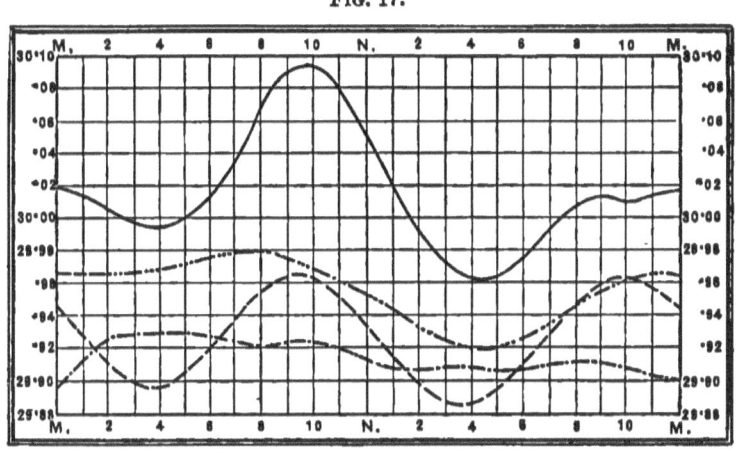

Diurnal Range of Pressure.
———————— Calcutta (January).
— — — — — Greenwich (January).
— ·· — ·· — Nertschinsk (May).
— · — · — · — Boothia Felix (January)

afternoon minimum differ respectively more from the mean than the other oscillations; or, in other words, the movement of the barometer is greater by day than it is by night. The difference between the extreme oscillations is termed *the diurnal range*, and this exceeds a tenth of an inch between the tropics, being 0·116 inch, on the mean of the year, at Calcutta, while in high latitudes its amount is far less, as appears from the curve for Boothia Felix. As Herschel says

('Meteorology,' p. 163), 'The diurnal oscillation is a phenomenon which invariably makes its appearance in every part of the world *where the alternation of day and night exists*. . . . Within the Arctic circle, however, the diurnal oscillation dies out, or rather merges in the annual.' These curves of variation can only be determined accurately by a long series of hourly or two-hourly observations, and accordingly they are only known for a comparatively small number of stations, so that it is hardly to be wondered at that the phenomenon has never been satisfactorily explained. About fifty years ago Dove proposed the subtraction, from the total barometrical pressure, of the tension of aqueous vapour (p. 110), as he found that by such separation of the total pressure into its two constituent parts—dry air pressure and vapour tension—the curve of the daily variation of the dry air was far simpler than the original curve, especially at continental stations. It will be seen from fig. 17 (p. 89) that at Nertschinsk in Siberia, where the air is very dry, the lesser maximum and minimum almost entirely disappear in May. This principle of separation is, however, not generally accepted at present, and we have not yet arrived at the interpretation of this regular barometrical motion.

Of late, however, a line of inquiry has been taken up which bids fair to throw some light on the cause of the phenomenon. Mr. Buchan and Mr. H. S. Eaton have independently investigated the problem of diurnal range in connection with the geographical position of the station where it is observed. Mr. Eaton has calculated the curves for the seven observatories in the British Isles for a year, and has shown how the continental character gradually imprints itself on the curves as we travel from the Atlantic coast towards the more continental stations (fig. 18). He has also shown that the difference in the barometrical curves is related to the daily range of temperature. Where the latter is small, as at Valencia or Falmouth, we have the minimum at 3 A.M. more marked than that at 3 P.M., while at Kew, where the thermometric range is comparatively great, the reverse is the case, and the afternoon minimum is the more important. Falmouth, however, as will be seen, ex-

Fig. 18.

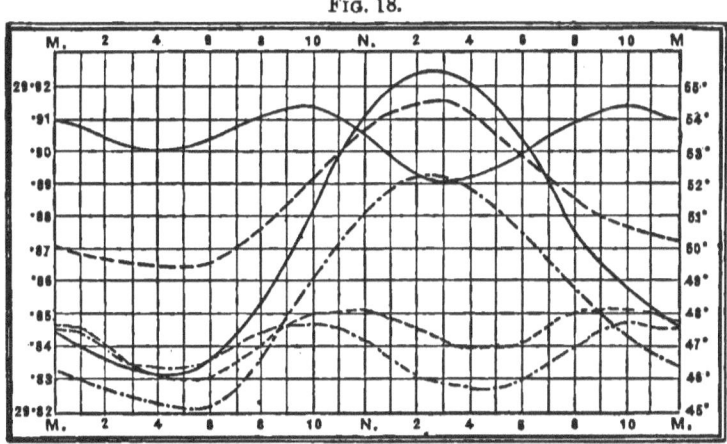

Diurnal Range of Pressure and Temperature (1876).
The temperature curves are those showing a single oscillation.
———— Kew.
—·—·—·— Falmouth.
— — — — Valencia.

hibits curves of pressure and temperature which are intermediate in character between those of Valencia and Kew. When we go to a really continental station like Nertschinsk, as has just been said, we find the morning minimum almost entirely obliterated in some months, and the curve apparently exhibiting a single oscillation.

Mr. Buchan's inquiry has been on a much more extensive scale than Mr. Eaton's, and he has not confined his discussions to stations for which the complete range-values have been calculated. Indeed, if he had, material would have failed him. In part of his investigations he has taken the simple difference between the forenoon maximum and the afternoon minimum, and he finds that the amount of daily amplitude so given is modified by proximity to the sea, or even to sheets of water like the great lakes of North America. This shows us that the amount of vapour present in the air is a factor which we must not disregard when dealing with this question, and yet our knowledge of the distribution of vapour in the atmosphere is very imperfect, as we shall learn in the next chapter.

In fact, when we begin to study the barometrical curves closely, we find that we want curves of equally minute accuracy for all the other elements for each station, before we can attempt to offer an explanation of the phenomenon.

In order to throw light on the geographical distribution of barometrical range, Mr. G. Harvey Simmonds calculated its elements for every station for which he could find a sufficiently continuous series of observations, taken at two-hourly intervals at least. The results were published by Mr. R. Strachan in the 'Quarterly Journal of the Meteorological Society,' vol. vi. p. 42. The list of stations in the table is, as might be expected, but small, only numbering thirty in all, but they cover a wide extent of latitude, from Hobarton in 42° 52' S. to St. Petersburg in 59° 57' N. The results, as far as they go, furnish evidence in support of Herschel's idea that diurnal range disappears in very high latitudes.

Fig. 19 exhibits the annual curve of pressure for London, for the equatorial part of the Atlantic, and for Central Siberia. It will be seen that, while the two former curves do not vary much from month to month, that for Barnaoul shows a very extensive reduction of pressure of more than 0·8 in. from winter to summer.

Barometrical pressure has a yearly as well as a daily range, and this annual curve presents a totally different appearance at continental stations from that which it exhibits, for instance, in these islands. All these changes are, however, connected with climate,

FIG. 19.

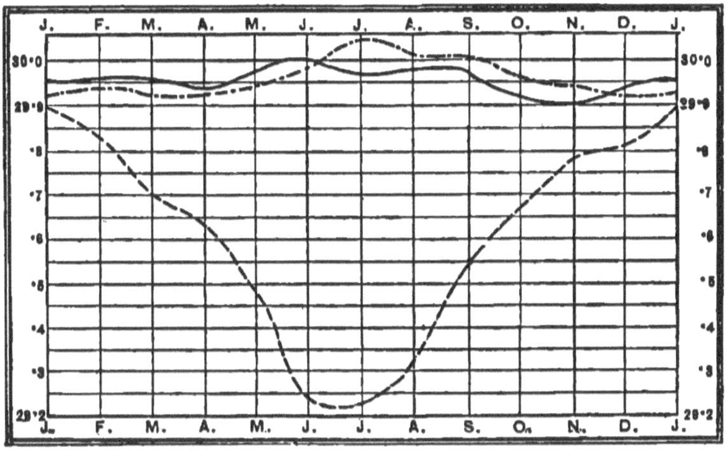

Annual Range of Pressure.
———— London.
—·—·—·— Equatorial part of Atlantic.
———— Barnaoul.

and with the seasonal changes in the distribution of the various meteorological elements. An examination of the charts of barometrical pressure for January and July, Plates VI. and VII., Chapter XIII., will give the reader a general idea of the regions where the barometer is highest and lowest respectively in each of these months, and of the extent of the annual change in different parts of the earth.

CHAPTER VI.

THE MOISTURE OF THE ATMOSPHERE.

WE have already said in Chapter II. that aqueous vapour is the most abundant element in the composition of the atmosphere, next to its principal constituents, nitrogen and oxygen. It is, moreover, the most important of all, meteorologically, inasmuch as it is that which, by its sensitiveness to the action of heat, and by its liability to change of condition from gaseous to liquid and solid, and *vice versâ*, is capable of exerting a great influence on almost all meteorological processes. The subject of this and the next chapter will be the determination of the amount of water passing into the atmosphere, present in it in the vaporous state, and finally passing out of it in a liquid or solid condition.

The subject, therefore, divides itself into three heads:—

1. *Atmometry*, or the determination of the amount of water passing into the air by evaporation.

2. *Hygrometry*, or the determination of the amount of water present in the air in a vaporous form.

3. *Hyetometry*, or the determination of the amount of water condensed out of the atmosphere in the form of rain, hail, or snow.

George Stephenson's saying that 'it is the sun which works railways,' is well known, and is perfectly

THE MOISTURE OF THE ATMOSPHERE. 95

true. It is equally true to say that it is the sun which gives us rain and all the good which rain brings us. Dove has said long ago that the atmosphere is a vast still, of which the sun is the furnace, and the sea the boiler, while the cool air of the upper atmosphere and of the temperate zones plays the part of the condenser; and we, on a wet day, catch some of the liquid which distils over.

1. *Atmometry* (or, more correctly, atmidometry, from ἀτμίς, 'vapour'), has, unfortunately, been far less studied than either of the other heads of the subject which we have mentioned; but, as we shall see, this is mainly due to the intrinsic difficulty of the observations to be made.

In order to explain the broad principles which govern the behaviour of the atmosphere with regard to moisture, we must state some general facts as to the mutual relation of gases and vapours.

A gas obeys the same laws, as to alteration of its volume according to changes of pressure and temperature, as hold good in the case of atmospheric air. A vapour obeys these laws *within certain limits*; but if these limits be overpast—*i.e.* if the pressure becomes too great or the temperature falls too low—a portion of the vapour will pass into the state of liquid. Under any circumstances of pressure and temperature, a given space can only contain a given quantity of vapour.

This is as true of vapour mixed with air as of vapour by itself. A given mass of air, under given conditions of temperature and pressure, can only contain a given quantity of water in the form of vapour; and if we attempt to reduce the temperature, or to increase the pressure, when the air is charged with vapour, a certain quantity of that vapour will be condensed and will pass into the liquid or solid form.

On the other hand, if a mass of air be not charged with vapour, and water be introduced into it in greater quantity than is required to saturate it, this water will evaporate until the amount present in the atmosphere in a vaporous condition is the greatest amount possible under the circumstances; or in other words, until the mass of air is charged with vapour.

Vapour which is just ready to be condensed is said to be at its maximum density, because any increase of density will produce condensation, and the air containing vapour in this condition is said to be *saturated*. If air be not saturated, as we have already said, it will continue to absorb moisture, or to cause evaporation, from any free water surface until it is saturated.

Moreover, vapour is given off from ice or snow as well as from water. If, during a hard frost, we counterbalance a piece of ice in a pair of scales, we shall see the scale containing the ice gradually rise, owing to the evaporation from the ice. We have all of us noticed the gradual disappearance of snow during frost, without any perceptible thaw taking place. In fact, in this case the water passes as rapidly from the liquid to the gaseous state as it has done from the solid to the liquid.

This process of water passing into the state of vapour is evaporation, and the converse process, of vapour passing into the state of water, is condensation. It is therefore pretty clear that, as these two processes are constantly going on, the amount of vapour actually present in the air is constantly changing.

The immediate effect of evaporation is to lower the temperature of all bodies in the immediate vicinity of the spot where the process is taking place. The reason of this is that as a large quantity of heat (the latent heat of evaporation) is required to convert the water into vapour, it is necessary that this heat should be available, and the only possible sources for it are the surrounding bodies. We therefore learn the reason why in summer wet cloths are wrapped round articles we wish to cool, such as bottles of wine. The evaporation of the water lowers the temperature of the bottle and its contents. Similarly, as we shall shortly see, the evaporation from the coating of the wet-bulb thermometer lowers the temperature of that thermometer. The loss of temperature over a given surface is of course proportional to the amount of water evaporated over that surface, so that the drier the soil the less is the fall of temperature produced by evaporation from it because the less is the amount of water present in it and available for evaporation.

The measurement of evaporation is a very uncertain observation, for since vapour behaves as a gas it is perfectly elastic and diffuses itself in all directions. If, therefore, we make an artificial pool of water, say in a perfectly dry field, the vapour rising from that pool will tend to diffuse itself through the surrounding atmosphere, and the rate at which that atmosphere will become saturated will depend on the extent of water surface supplying it with vapour. Hence we see that if we measure the evaporation by a vessel of manageable size, we can never be sure that these indications bear any direct relation to the rate at which evaporation takes place from a lake or a pond, or from the surface of the ground itself, which is always more or less moist.

Again, if we freely expose a vessel, say a flat dish, to the air for any considerable time, it will receive rain, so that we must ascertain how much rain has been caught in the dish in order to be able to calculate how much water it has lost by evaporation.

If we want to avoid this source of complication we must put the atmometer under a cover, say in an open shed, but then we do not secure a free exposure of the instrument to the sun and air.

Lastly, the management of the apparatus requires special care. If the evaporating surface be large the action of the wind may cause ripples, and thereby splashing and loss of water. If the surface be small the level of the liquid must be kept, as nearly as possible, flush with the rim of the dish, for the least depression below that rim produces a serious diminution in the amount of water vaporised, for the reason shortly to be explained.

Supposing the air to be quite dry, the amount of evaporation from a water surface exposed to it is regulated (a) by the temperature—for the higher the temperature rises, the more rapidly is the water converted into vapour, the more vapour will the air contain; and (b) by the wind—for the more promptly the air in contact with the water is changed, the more quickly will that water give off vapour to it, because the vapour is removed as fast as it is formed.

Hence we can at once see that if the level of the water in an evaporating pan sinks below the rim of the dish, there will be a layer of still air beneath that rim, resting on the water, which will not be removed by the wind as easily as the layer flush with the rim.

It will be seen from what has already been stated, that the difficulties of the observation are serious, but to give some idea of its practical importance to hydraulic engineers, irrespective of its purely scientific interest, we may add that it has been calculated by Mr. A. R. Binnie that from a reservoir at Nagpoor, in India, during the dry season of 240 days, as much as four feet of water, or one-fifth of an inch per day, were evaporated, and the entire loss due to this cause was 54 per cent., or more than one-half of the annual supply to the tank. We shall see presently that this result is not at all extraordinary for the locality where it was obtained.

Another very important practical application of atmometry is to the determination of the rate of drying of the soil, for this is a process which exerts a most important influence on the growth of plants. Here, however, the nature of the crop has to be taken into consideration. It has been proved by Sir J. B. Lawes and Dr. Gilbert at Rothamsted, that a deep-rooted crop, like wheat, will thoroughly dry the ground to a depth at which, in similar circumstances of weather, the store of water under a grass meadow would be unimpaired.

There is another remark which presents some practical interest to those who have to deal with plantations and pleasure grounds. It is evident that the rate of evaporation from bare ground is much more rapid than from earth covered with leaves, &c. If, then, the leaves be removed in autumn, we not only deprive the soil of the salts which should be restored to it by the gradual decay of the foliage which it has borne throughout the summer, but we also abstract from it much of the moisture which otherwise would have rested about the roots in winter time. These are, however, digressions, and merely indicate some of the uses to which a knowledge of evaporation may be applied.

We have already explained that few atmometers give really satisfactory results. Those of small size are usually exposed in thermometer screens, and therefore more or less sheltered from the action of the sun and air.

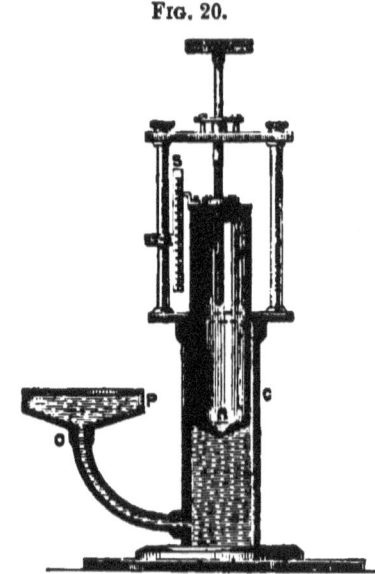

Fig. 20.

Von Lamont's Atmometer.

Professor von Lamont's apparatus, which may serve as an example of smaller instruments, is shown in fig. 20. It consists of a pan P, connected by a pipe with a cylinder C in which a piston plunger R, carrying a scale S, is moved up and down by a screw. The scale is set at zero, and then water is poured into the pan until it stands exactly flush with the opening of the outflow pipe O. The piston is then screwed down till the level of the water is raised to the rim of the pan, and the apparatus is left to itself. When the time of observation comes, the piston is screwed up again till the level of the water sinks to the opening of the outflow pipe, and the reading on the scale, showing the change of level of the piston, shows the amount of evaporation.

Other instruments, which may also be employed for the measurement of evaporation from earth as well as from a free water surface, are made on the principle of a balance, and measure the loss of weight in a given time.

As regards the results of evaporation experiments, such is the discordance in the results obtained, as already pointed out, that Schmid, in his 'Lehrbuch der Meteorologie' (1860), says at page 600; 'The final result of all the observations on evaporation, and these are very few, can be easily given in the statement that it is utterly impossible to estimate, even approximately, the quantity of water which passes into the atmosphere at a given place and in a given time.'

This is rather disheartening, but since the date of Schmid's book attempts have been made to determine, with some success, the amount of evaporation from a free water surface exposed to the sun and wind.

The most complete discussion of this subject as yet accessible will be found in Haughton's 'Six Lectures on Physical Geography' (Longmans, 1880), p. 165. For the torrid zone and its immediate vicinity the following data exist:—

Locality	Latitude	Evaporation
Madras	13° N.	91·25 inches.
St. Helena	17° S.	83·78 ,,
Nagpoor	29° N.	73·15 ,,

The evaporation at the equator is deduced from these values by the formula

$$e = a \cos \lambda,$$

where e is the evaporation at a station at the latitude λ, and a that at the equator.

This formula gives the mean evaporation at the equator as 88·23 in. Now, as only three-fourths of the surface of the globe at the equator is covered by water, we may reduce this amount by one-fourth, and it becomes 66·15 in. for the total evaporation, while the average amount of rainfall for the equator is estimated by Dr. Haughton at 66·84 inches.

The following observations for French stations, given by Cole in 'Atmidométrie' (Orléans, 1866), lead to a similar result:—

Locality	Rainfall	Evaporation
St. Jean de Loane, sur la Saône	30·8	25·9 inches.
Dijon	27·7	26·2 ,,
Pouilly	30·4	22·4 ,,
Montbard	27·2	23·2 ,,
Laroche-sur-Yonne	22·4	21·7 ,,

For these islands, Mr. C. Greaves, C.E.,[1] found for London, on the average of fourteen years, 1860–1873, the rainfall 25·721 in., and the evaporation 20·613 inches. In only three years did the evaporation exceed the rainfall.

Lastly, Dr. Haughton himself found for Dublin on the average of two years that the evaporation fell short of the rainfall by 1·08 inch.

These figures therefore show that probably, in nearly all parts of the globe, situated reasonably near the coast, the rainfall is about equal to the evaporation from a *free water surface*, and that there can be no great transference of vapour from the torrid to the temperate zones. At times the balance may incline one way and at times the other: the rainfall may exceed the evaporation for a lengthened period, and again for

[1] *Minutes of Proceedings of Inst. C.E.*, vol. xlv.

months together not a drop of rain may fall; but the general result on the whole year is as given above.

2. *Hygrometry* (from ὑγρός, 'moist') is the determination of the amount of moisture present in the air in a vaporous form. Of hygrometers there are two principal divisions—Direct and Indirect; and these latter are further subdivided into two classes—Organic and Inorganic.

Direct Hygrometers.—The simplest of all hygrometrical experiments is made in summer, whenever a glass of cold water is brought into a warm room and drops of water appear on the outside of the glass. The reason of this is, that if the water is at a temperature below that corresponding to the tension of the vapour present in the atmosphere of the room, moisture will be condensed on the vessel containing it. As soon, however, as the water rises above that temperature this process of condensation ceases.

Of direct hygrometers, one of the oldest is Daniell's, and among the most modern and simplest is Dines's. They all depend on the principle that the moment the temperature of any object falls below the temperature corresponding to the maximum tension of the vapour present in the atmosphere at the time, there will be a deposition of moisture upon it, which will at once be visible if the surface be of polished metal or of glass. The temperature at which this deposition takes place is called the Dew-point, as we shall see in the next chapter.

In Daniell's instrument we have two bulbs connected by a tube bent twice at right angles. One bulb is of black glass, the other is of ordinary glass, but is coated with linen. A thermometer is enclosed in the instrument, with its reservoir in the black bulb, of which it therefore shows the temperature. A certain quantity of ether is contained in the instrument. The mode of using the apparatus is to drop a little ether on the linen coating of the clear bulb. This by its evaporation makes the temperature of that bulb to fall, and causes the ether inside the instrument to distil over from the black bulb, which in its turn falls in temperature, owing to the evaporation of the contained ether, and eventually becomes coated with moisture. The instant at which the moisture first appears on the surface of the black bulb is that at which the temperature of the enclosed thermometer should be read. This temperature is of course the dew-point.

It is not an easy matter to make an observation with this instrument, and the necessity of employing ether is a serious inconvenience. It is practically never used in meteorological observations in recent times. Regnault's hygrometer is an improvement on Daniell's, as regards facility of manipulation and certainty of observation; in it the deposition of moisture takes place on a bright silver cup, which betrays the slightest trace of dulness, and the ether vapour is withdrawn as fast as it is formed, by the use of an air pump, or of an aspirator, so that the evaporation is accelerated.

Dines's instrument is very simple (figs. 21A and 21B), consisting of a vase A fitted with a pipe at the bottom, which is conducted close under a plate of black glass, where it also envelops the bulb of a thermometer C; a cock B is fitted at the base of the vase. Very cold water, or ice and water, is put into the vase, and the cock is opened; the glass speedily becomes dulled, and the thermometer is read. The cock is then closed again, the water in the tube soon rises in temperature and the cloud disappears, the moment of its disappearance being that when the dew-point is again reached.

Fig. 21 A

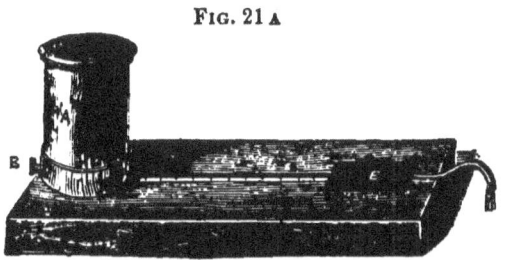

Dines's Hygrometer.

The operation may be repeated as long as the water in the vase remains at a temperature below the dew-point. This form of hygrometer is very well suited for use in any rooms where it is required to regulate the dampness of the air very carefully.

Fig. 21 B.

Dines's Hygrometer.

Indirect Hygrometers.—We shall first treat of those which are organic. These all depend on the well-known property of organic bodies to alter their molecular arrangement or their appearance according to their hygrometrical condition. We all know that ropes become shortened by wet, but hair is affected in the opposite way, and grows longer when it grows damp. Saussure was the first to adapt this principle to scientific uses.

The essential part of his instrument—the Hair

Hygrometer—is a human hair which has never been oiled or roughly handled. Such a hair stretches as it grows damp and contracts as it dries. It is kept fixed at one end and stretched by a light weight at the other, the cord of connection passing round the sheaf of a block, to which is attached an arm moving on a graduated arc. The position of the arm on the arc, therefore, depends on the length of the hair from its point of suspension above.

It is obvious that, though such an arrangement can be made to indicate the amount of moisture in the air, it cannot afford an absolute measure of that moisture, inasmuch as the normal length which the hair possesses, when the atmosphere is charged with moisture, must be determined for each hair individually. This is effected by observing the instrument at a time of perfect saturation of the air, produced either naturally or artificially, and moving the upper screw till the index is brought exactly to 100—the point of perfect saturation. The instrument is then fit for use.

I have explained the principle of the hair hygrometer at some length, as, although it is seldom used in this country, it affords practically the most satisfactory means of determining the hygrometrical condition of the air at temperatures close to and below the freezing-point, under which circumstances the dry-and-wet-bulb hygrometer fails.

Numerous other attempts have been made to utilise organic substances for hygrometrical observations, but none of them have resulted in the production of a useful scientific instrument. As an instance we may cite the use of seaweed, which, owing to the hygroscopic properties of the salts which it contains, grows

THE MOISTURE OF THE ATMOSPHERE. 107

damp in wet weather and dries when the weather clears up. The well-known weather-house toy, in which a woman comes out when the air is dry and a man when it is damp, depends on the expansion and contraction of a piece of catgut, according to the hygrometrical condition of the atmosphere.

Of *Indirect Inorganic Hygrometers* the only one to be noticed is the dry-and-wet-bulb hygrometer, first proposed by Sir John Leslie in Edinburgh,[1] and subsequently by Mason in this country. This is the apparatus almost universally employed at present, and consists of two thermometers, of which one has its bulb coated with muslin (fig. 22) and kept moistened with water. The principle of the instrument is that as long as the atmosphere is not saturated with moisture evaporation will take place from any damp surface exposed to it, such as the moist coating of the wet bulb. If the air be saturated no evaporation is possible, and the two thermometers will read alike; but if it be not saturated the temperature of the coated bulb will fall until it reaches a certain point intermediate between the temperature of the dry-bulb thermometer and of the dew-point. Once that limit is reached, if the supply of water be kept constant, the wet bulb will not change its indication unless the actual amount of moisture in the air varies.

FIG. 22.

Wet Bulb Mounting.

The supply of water to the wet bulb is effected either by moistening it with a wet brush, or, for con-

[1] Hutton in 1792 had noticed that a thermometer read lower if the surface of the bulb was wet.

tinuous observations, by wrapping a few threads round the neck of the bulb (fig. 22), and allowing them to dip into a water-vessel placed at a distance of a few inches. Such an arrangement must fail in frost, as then capillary action along the threads will cease; and also care must be taken to make the supply more copious in very dry weather than at other times. Full directions for managing the wet bulb thermometer will be found in the 'Instructions' already quoted.

The theory of the instrument has been investigated in Germany by August and in this country by Professor Apjohn ('Transactions of the Royal Irish Academy,' vol. xvii.), who gives the following equations:—

$$f'' = f' - 0{\cdot}1147\,(t - t')\,\frac{p - f'}{30},$$

where f'' = the tension of aqueous vapour existing in the air, the temperature corresponding to which is the dew-point.

f' = the tension of vapour corresponding to the reading of the wet bulb.

t = the temperature of the dry bulb.

t' = ,, ,, wet bulb.

p = the reading of the barometer.

The fraction $\frac{p - f'}{30}$ usually does not differ very much from unity, at stations situated near the sea level, and the formula is commonly abbreviated thus:—

$$f'' = f' - \frac{t - t'}{87}.$$

For temperatures below the freezing-point this formula becomes—

$$f'' = f' - \frac{t - t'}{96}.$$

These formulæ, strictly speaking, require the use of tables of the tension of aqueous vapour, but, for convenience' sake, special tables have been constructed, giving the dew-point and other hygrometrical results by inspection. The tables most generally used in this country are Glaisher's,[1] and are based on a series of numbers called the Greenwich Factors.

These factors were determined by comparisons between simultaneous observations of the dry and wet bulb hygrometer and Daniell's hygrometer, made at Greenwich, with similar comparisons of observations taken at high temperatures in India, and others made at low and medium temperatures at Toronto. The tables are fairly accurate, for the conditions and range of humidity which usually occur in the neighbourhood of London; but they are insufficient for the conditions of great dryness which occasionally are noticed here, and, à fortiori, they fail in the extreme aridity of the climate of India, or in the intense cold which occurs, for instance, in Canada. That this must be so is evident when we point out that the greatest range of temperature for which provision is made in the factors is from 10° to 100°, and that the greatest depression of the wet bulb, below the dry, allowed for in the tables is 24°—limits insufficient to meet the requirements of such climates as have been mentioned.

The greatest objection to the wet bulb hygrometer is the difficulty of managing it in frost. Although a formula has just been given for readings below 32°, it is evident, from the sudden change in the denominator

[1] *Hygrometrical Tables, adapted to the use of the Dry and Wet Bulb Thermometer.* By James Glaisher, F.R.S. London: Taylor & Francis, 6th edition. 1870.

of the fraction, that there is some uncertainty in the results just at the freezing-point, and it has been found practically impossible to secure a satisfactory continuous record during frost. Under such circumstances the use of the hair hygrometer has been enjoined in Russia and elsewhere.

The results of hygrometrical observations may be calculated in various ways.

Firstly, we can determine the Dew-point, as has already been explained.

Secondly, we can determine the actual pressure corresponding to the amount of vapour present in the air at the time. This is called the Vapour Tension, or sometimes, but incorrectly, the Absolute Humidity. This result may, by an easy calculation, be expressed in terms of the weight of water contained in a given volume of air, *e.g.* in a cubic foot.

Thirdly, we can ascertain from tables the greatest amount of vapour which the air could possibly contain at its temperature at the time of observation. This is given by the vapour tension corresponding to the temperature of the air. If the vapour tension given by direct observation corresponds with this amount, the air is said to be saturated, but if not, the proportion which the actual vapour tension bears to this maximum amount is called the Fraction of Saturation, or, more commonly, the Relative Humidity.

We now come to consider the part which aqueous vapour plays in the atmosphere; and here we are met by the same great difficulty which we have noticed in dealing with other elements. We can only determine the amount of vapour in the portion of air actually in contact with the instruments, and, from what we have said about evaporation, it will be seen that the wind will exercise a very great influence on all hygrometrical observations, and especially on those of the wet bulb hygrometer.

It is also very probable that the distribution of vapour is extremely local, depending on the proximity of free water surfaces to supply it to the atmosphere; but in general terms it may be said that the curves of vapour-pressure over the globe, as laid down by Professor Mohn,[1] exhibit a striking concordance in their course with those of temperature.

The distribution of vapour-pressure in vertical height has a most important influence on the condition of the atmosphere, and about this our knowledge is very scanty. In general terms we may say that the vapour, like the temperature, decreases as we ascend, but the rate of diminution is much more rapid in the former case than in the latter.

Kämtz showed that, in September and October 1832, the vapour-pressure at Zürich was 0·366 in.; while at the Faulhorn, at the height of 8,400 feet above the sea, it was only 0·162 in.; and this proportion was kept up in the succeeding year, which was a wet one, while 1832 was dry. It has been calculated that one-half the quantity of vapour in the air is contained in the lowest 6,000 feet of the atmosphere, and that the amount contained above the level of 20,000 feet is only one-tenth of that at the surface of the ground. In fact, the vapour in the atmosphere decreases much more rapidly than would be the case in an independent vapour

[1] *Grundzüge der Meteorologie.* H. Mohn. 2nd ed. Berlin: D. Reimer. 1879, p. 97.

atmosphere existing under its own pressure, as has been shown by Lieut.-Gen. R. Strachey ('Proceedings of the Royal Society,' vol. xi., p. 182). This diminution of vapour-pressure does not proceed at a uniform rate, for in Daniell's 'Meteorology'[1] we have an account of an ascent of the Peak of Teneriffe, by Captain Basil Hall, who found that he passed through successive strata of vapour—'vapour-planes,' as Herschel has called them. After getting wet through at Orotava Captain Hall found the air very dry at the summit of the mountain, but on descending he re-entered the damp stratum at the lower level. This experiment has unfortunately not been systematically repeated, but it is confirmed by occasional observations in balloons. These considerations are, however, sufficient to show that we are not safe in drawing precise conclusions as to the amount of vapour in the atmosphere at large from isolated hygrometrical observations taken at the surface of the earth. But, in a general way, it has been inferred by Strachey from observations, and by von Lamont from theoretical considerations, that the actual tension of the vapour existing in the atmosphere is only about one-fourth of that observed at the surface of the ground.

This shows us that Dalton's law, that gases mixed together exert each its own pressure independently of the others, is not applicable to the mixture of dry air and aqueous vapour in the atmosphere, although the law is strictly true in the case of a confined portion of damp air, such as may be made the subject of a physical experiment.

The reason of this difference is that the relations on

[1] Daniell's *Meteorological Observations and Essays*, 2nd edit. 1827, p. 593.

which the law is based can only hold good when the air and vapour are uniformly mixed throughout the entire extent of the atmosphere—a condition which, as we have just seen, is not fulfilled in nature.

From the foregoing statements we learn that it is incorrect to assume that the amount of vapour-tension at the earth's surface may simply be subtracted from the barometrical reading in order to obtain the pressure of the dry air alone. Accordingly, Dove's theory of the independent actions of dry air and of aqueous vapour (p. 90) has not met with general acceptance in recent times, and has been completely abandoned in modern publications.

The principal action of the vapour in the air, so long as it remains in the vaporous condition, is as a medium which restricts the access of heat to the earth from the sun, and also the radiation of heat from the earth into space. Tyndall expresses this by saying that a sheet of vapour acts as a screen to the earth, being in a great measure impervious to heat. The intensity of solar radiation on high table-lands and in the polar regions, as contrasted with that experienced under other circumstances, which we have noticed when treating of radiation, is chiefly due to the comparative dryness of the air. It is the invisible vapour in the atmosphere which checks the excessive loss of heat by radiation from the earth on a clear night; while every one knows that, if the moisture is present in abundance, in the form of dense clouds, the loss of heat by radiation is very slight indeed.

CHAPTER VII.

PRECIPITATION—DEW, FOG, MIST, AND CLOUD.

WE now come to consider the way in which the water which has passed into the atmosphere in a vaporous state is again restored to the earth in the form of dew or rain. This branch of the subject is called *Hyetometry*, from ὑετός, rain.

Dew is the precipitation of moisture without the formation of visible cloud. In the Middle Ages it was even considered a product of the stars, and was collected by the alchemists. The true explanation of the phenomenon was given by W. C. Wells, a London physician, who died in 1817. He found that if he exposed a flock of wool on the ground when dew was forming, moisture would be deposited on the wool if it was freely exposed to the sky; but that if it was placed under a table, or under any covering, no moisture would be found on it.

The process of the formation of dew is as follows: When in the evening the heat received by the earth from the sun begins to fall short of the heat given out by the earth to space, the temperature of the earth's surface naturally falls. If the sky be clouded, most of the heat radiated from the earth will be reflected back; but if the sky be clear, especially when the atmosphere is rather dry, the chilling of the ground may be very considerable. When the surface of the earth becomes colder, the air in contact with it is cooled also. If this cooling process lowers the temperature of any portion of the air till it reaches the point for which the corresponding vapour-tension is that of the vapour actually present in the air, this vapour will be on the verge of condensation, and the slightest further reduction of temperature will cause the deposition of dew. This point is therefore called the Dew-point. If the dew-point be below the freezing-point, the moisture deposited will pass at once into the solid form, and we shall have the phenomenon of hoar-frost. This is not frozen dew, but water deposited in the solid form. It is, of course, deposited most copiously on the objects which are the best radiators, and will therefore be found on grass and herbage. Trees are especially prone to exhibit an abundant coating of hoar-frost, forming most lovely objects on a winter's morning. On a dry road the twigs, &c., which are lying about, will bristle with spiculæ of ice, while the ground about may show not a trace of the deposit.

I may here refer to the phenomenon of 'silver thaw,' or 'glazed frost' (Fr. *verglas*, Germ. *Glatteis*), which is frequently, though incorrectly, considered to be of the nature of hoar-frost. The glazed frost is really the frozen surface which is occasionally produced at the beginning of a thaw, if a warm wind suddenly sets in. The damp air, passing over the ground, of which the temperature is exceedingly low, has its moisture deposited in the solid form, and all objects on which this deposit takes place are covered with a sheet of ice. This phenomenon is intensified if a fall of rain occurs at the time.

One of the most striking instances of glazed frost, and of the damage which it can occasion, occurred in France, January 22–4, 1879, in the department of Loiret, and also in the neighbourhood of Paris, and was described by M. Godefroy in the 'Comptes Rendus'[1] of the French Academy for the time. He says that 'rain fell continuously for the three days, although the air temperature ranged from 24° to 28°. When the rain was scanty, each drop at once solidified even on warm objects, and took the form of small, flattened, irregular pastilles. The phenomenon was especially remarkable when the rain fell on woollen stuffs. The drops had evidently been brought to a state of *superfusion* in their passage through cold air; that is, they had been cooled below the freezing-point without congealing, so that they at once solidified on contact with solid bodies. When the rain was plentiful, on the other hand, part of it was at once changed to ice, but part remained for a time in the liquid state, flowing along solid bodies, forming new layers of ice, and producing stalactites. The ice-covered branches of trees broke more and more under the weight, and on the evening of the second day the phenomenon assumed frightful proportions. In the morning the ground was strewn with branches, whole trees lay uprooted, and others were entirely split from top to base. Most of them were entirely cleared of branches, and in some parts the forest looked like one of masts. The following statements will explain the destructive action. A twig of a lime tree 4 inches long weighed, when covered with ice, 930 grains; the same twig, when freed from ice, weighed 7·5 grains. A leaf of laurel carried a coating

[1] *Comptes Rendus*, vol. lxxxix., p. 999.

of ice weighing 1,120 grains. All objects exposed were alike covered with ice. An evergreen shrub, like a rhododendron or a privet, became a block of ice, through which the leaves and branches could be distinguished fairly well. Fir trees had the appearance of huge pyramids of ice, each group of branches being weighed down on the one below, and the lowest on the ground.'

To return to the subject of hoar-frost. The process of condensation of vapour sets free a large amount of latent heat, and this has a tendency to retard the cooling of the atmosphere, so that, during the night, the temperature of the earth cannot easily fall below the dew-point which existed at nightfall; for if it should fall below that point a more copious condensation of moisture would of course set free more latent heat, and send up the temperature above the dew-point again.

These facts, as Buchan points out, indicate a very practical use of hygrometrical observations to gardeners and others, for a knowledge of the dew-point in the evening may enable them to form a fair estimate of the probable minimum temperature on the grass during the night. If the dew-point be found to be above 32°, they need not fear that hoar-frost will occur; but if it be below 32°, it is advisable to cover up delicate plants, for fear of their being nipped if the sky should clear before sunrise. It is known to every one that dew does not appear on a cloudy night, and in some places artificial clouds are produced, by burning weeds and other substances which give off a copious supply of smoke, and are intended to form a layer over the crops which it is desired to protect from the deleterious influence of the frost. Trustworthy evidence of the real efficacy of this process is very desirable. The action is supposed to be similar to that of the table in Wells's experiments.

Dew can never appear when there is much wind, for the air cannot remain long enough in contact with the soil for any material reduction of its temperature, and consequent condensation of its moisture, to take place.

The cooling effect of radiation, and the resulting formation of dew, are most active when the radiating power of the substance with which the surface of the ground is covered is greatest. The radiating power is greatest where the vegetation is rankest, and so while on a road, on a gravel walk, or in a paved courtyard, no perceptible moisture may be found, the surface of a meadow close at hand may exhibit an abundant precipitation of dew.

In some of the forests of South America, according to Humboldt, the traveller on entering a wood finds apparently a heavy shower falling, while outside the wood the sky is perfectly clear. Here the formation of dew takes place at the tops of the trees, and so copiously, owing to the abundance of vapour in a tropical atmosphere, that a real shower of rain is the result.

It is, however, clear from the foregoing that any attempt to measure the amount of dew quantitatively must be very difficult. However, Mr. G. Dines, F.M.S., in a paper 'On Dew, Mist, and Fog,'[1] gives an account of a series of experiments which he has made to determine the amount of dew; and he says, 'Taking

---

[1] *Q. J. Met. Soc.*, vol. v., p. 157.

the averages of all my observations, we get the value of 1·397 inches. If the observations on grass only are taken, the amount obtained is 1·022 inches. Looking at these figures, and making a liberal allowance for contingencies, it may, I think, be fairly assumed that the average annual deposit of dew on the surface of the earth falls short of 1·5 inches.'

The greatest amount of the water which passes into the atmosphere by evaporation is restored to the earth in the form of rain, snow, or hail, but before the moisture takes the form of drops of appreciable size it passes through the intermediate condition of visible fog, mist, or cloud.

We know that aqueous vapour is as perfectly transparent as air or water. It is only when the water is disseminated through the air, not as vapour, but as liquid in a state of minute subdivision, that the mixture becomes opaque; just as a transparent sheet of glass, when pounded up, becomes an opaque white powder, owing to the intermixture of particles of air and glass, with their very different refracting powers.

The first step in the conversion of aqueous vapour into rain is the production of extremely small droplets, the form in which water is present in fog or cloud, or in the steam cloud from a locomotive. This visible steam is not vapour at all, but water in a state of very minute subdivision. As soon as it has been produced, the process of condensation is complete, and all the latent heat which is to be liberated has been set free. The particular agency which causes these very small droplets to coalesce into rain-drops is not yet understood, and much less has the production of large hailstones been fully explained, but there is reason to think that electricity plays a great part in these processes.

Fogs are frequently caused by the intermixture of masses of air of very different temperatures. If the mixture of two bodies of air has a temperature lower than is requisite to maintain, in the form of vapour, the total quantity of moisture contained in the mass, a portion is, to speak chemically, precipitated, in the condition of minute particles of water, as already explained. The condensed moisture has a tendency to deposit itself on foreign particles floating in the air, and accordingly the fogs of London are pre-eminently opaque, owing to the enormous amount of carbon, in the form of soot, which the atmosphere holds in suspension. In fact, Dr. Aitken has recently published, in the 'Transactions of the Royal Society of Edinburgh,' vol. xxx., p. 337, a paper 'On Dust, Fogs, and Clouds,' intended to prove that, unless there be some nucleus, some foreign body, floating in the atmosphere, no fog or cloud can be produced by the condensation of aqueous vapour.

Fogs are, however, generated by the cooling of masses of air in other ways than by mixture. If a warm damp current of air flows over a chilled surface, such as that of an ice-floe, or a cold ocean current, a fog is the result, and the density of the fogs on the Banks of Newfoundland is well known, where they are often as sharply defined in extent as the hardest looking wool-pack clouds. It is not an uncommon phenomenon, when a fog-bank is lying off the harbour of St. John's, for the bowsprit of a ship to be seen emerging from the fog, while not a trace of the masts or hull is perceptible; and again at times the topmasts will be in bright sunshine while the crew cannot see from stem to stern on deck. In a similar way cold rivers, such as those which flow down from glaciers and pass through the warmer air lying over the open country, produce dense fogs.

A converse cause of the production of fogs is the sudden chilling of saturated air lying over a warm water surface. This is the origin of the fogs which form over running streams in this country in time of frost. The water being warmer than the air, gives off vapour in greater quantity than the air at its low temperature can contain, and the surplus amount is precipitated as fog.

A class of fogs, termed by Herschel 'radiation fogs,' are formed in valleys or over damp meadows in the evenings. The ground is cooled by radiation below the dew-point, and all the superfluous moisture in the lowest stratum of air is deposited in the form of dew, while the air close to the soil is left saturated, and at the same time colder than the strata above it. If now no mixture takes place, no fog will result, and therefore none is formed over plains, where the strata of air are likely to be in equilibrium; but if there be the least difference of level, the colder air on the upper ground will flow down and produce a disturbance of the arrangement of the air-layers below, causing a mixture, which results in the condensation of a quantity of moisture in the form of fog. We often see such fogs over low grassy bottoms, or filling valleys to a certain height, with a surface as level as a lake.

These fogs are produced by cold air flowing down into a warm and damp atmosphere, and they disappear after sunrise, as soon as the heat is sufficient to enable the air to take up all the moisture which has been condensed out of it.

Similar in their nature are the sea fogs so common on our coasts in summer, which do not extend more than a mile or two inland. The air over the sea is charged with moisture, and when a cold wind blows over it, fog is the result, which may be carried some little way inland, but cannot be produced over the surface of the ground on shore, because the air in contact therewith is drier.

A most instructive illustration of the formation of fog has been given me by Dr. Wijkander, from his experience in the Swedish Expedition to Spitzbergen in 1872. The vessel was in winter quarters in Mossel Bay, not far from the north point of Spitzbergen. The country is exceedingly mountainous, and the land between Mossel Bay and Hinlopen Strait is high. The strait runs nearly due north and south. The eastern coast of Spitzbergen is constantly ice-bound, while the west and north coasts have open water, for they feel the influence of the Gulf Stream. The southerly wind blowing from the ice surface on the south-east side along this strait, the only channel of connection, generates a dense fog as soon as it comes in contact with the comparatively warm air over the open water to the northward, while it cannot so easily cross the mountains and descend on Mossel Bay. Accordingly we can easily understand how the expedition came to see a perpetual fog hanging over the northern entrance to Hinlopen Strait, while they themselves, at a few miles distance, were in bright sunshine.

Mist is very similar to fog, but the particles of moisture are somewhat larger, and, as is well known, a mist is much wetter to the feel than a fog. Clouds, however, are really nothing else but fog or mist, and the most solid looking night-cap on a hill-top is found by those who are enveloped in it to be neither more nor less than driving mist.

The reason that clouds form pre-eminently on mountain-tops is that the upper parts of mountains are cold, and so the warm lower stratum of air, forced to ascend in order to pass over them, is chilled and its moisture condensed; but as soon as a body of air in its course has passed over the mountain top, and has thus got outside the sphere of the chilling influence of that cold body, the contained moisture can return to the state of vapour, and become as invisible as before.

Clouds, however, are not always composed of water. There is an all but absolute certainty that the clouds belonging to the higher stratum, which we shall presently describe as upper clouds, are really composed of spiculæ or small needles of ice. This is shown by the fact that at the level at which such clouds float the temperature must be so low that no water could exist in the state of liquid; but a more direct proof is afforded by the fact, which we shall shortly consider, that halos and mock suns, &c., are never formed by any but upper clouds, and the origin of these phenomena cannot be explained in accordance with optical laws, except on the hypothesis that the light is refracted through prisms of ice.

The level at which the lowest clouds float is, in South America, about 9,000 feet; in the Tyrol it sinks to about 5,000, and in these islands, out of a great number of observed cloud levels, one-third were below 2,500 feet.

Clouds were first systematically studied by Luke Howard,[1] and his classification of them has met with general acceptance hitherto, no attempt to remodel it having been received with much favour. Howard recognised three primary, with four compound, types, the latter being the combinations of the simple forms.

The primary types are—
A. *Cirrus*. The mare's-tail cloud.
B. *Stratus*. The ground fog.
C. *Cumulus*. The wool-pack cloud.
D. By the combination of A. and B. we get the *cirro-stratus*.
E. By the combination of A. and C. we get the *cirro-cumulus*.
F. By the combination of B. and C. we get the *cumulo-stratus*.
G. By the combination of all three primary types we get the *cumulo-cirro-stratus* or *nimbus*, the rain-cloud.

In order to consider these systematically we shall classify them under the two categories of upper and lower clouds. The types which belong to the upper series are A, D and E.

Cirrus is the streaky cloud, like a feather or a spray, which appears at a very great height in the atmosphere, far above the highest mountain peaks. It is generally a sign of wind.

[1] *Essay on the Modifications of Clouds*, by Luke Howard, F.R.S. [First published 1803.] 3rd edition, 1865. London: John Churchill.

Cirrus usually moves in a direction differing from that of the wind at the surface of the earth, but its motion often appears to be so slow that there is great difficulty in determining it without watching for a considerable time, so as to mark the motion of the cloud over some fixed object. However, the importance of the observation makes it highly desirable that particular attention should be devoted to it. The subject has been studied in this country by the Rev. W. Clement Ley, and in Sweden by Professor Hildebrandsson of Upsala, and from their investigations most interesting conclusions have been drawn as to the movements of the atmosphere in regions which, from their height, are quite inaccessible to any other class of observation. A fuller account of these investigations would, however, more properly belong to a book dealing with weather and weather prediction.

Cirro-stratus is usually generated by increased condensation in the cirrus already formed, which consequently sinks to a lower level.

The first sign of a change of weather from a clear sky is usually the development of cirrus. This gradually spreads its network over the sky; the fibres then grow coarser, and at last coalesce to form a grey film. At times the ribs of cirro-stratus stretch from one point of the horizon to the opposite point, the lines, owing to the effect of perspective, appearing widest apart at the zenith, and converging on either side like the streaks of a boat. This phenomenon is very common, and is known in some parts as a 'Noah's Ark' in the clouds. It is an almost unfailing sign of rain. When bad weather is approaching, the cirro-stratus cloud increases in compactness and density, and sinks to a lower level, at times entirely intercepting the direct rays of the sun and moon, and presenting the appearance of a uniform sheet overspreading the sky.

It is in cirro-stratus that the optical phenomena, which we have already mentioned, halos, mock suns and mock moons, are manifested, and these prove that cirro-stratus is an ice-cloud. The more common appearances of *coronæ*, rings or burrs round the moon, are produced by droplets of water, and appear when any thin cloud crosses the moon.

For the uniform sheets of which we have spoken, M. Poëy has proposed the name of *Pallium*, a cloak, but this term has not met with general acceptance.

Cirro-cumulus is also a high cloud, though usually lying at a lower level than the cirrus, from which it differs in being more globular in form, and consisting of small detached rounded masses, like a flock of sheep lying down, or like the markings of a mackerel, whence the name 'mackerel sky.'

Before leaving the subject of upper clouds altogether it should be remarked that as clouds are seen at all levels between the highest cirrus and the lowest stratus, it is often difficult to determine whether a particular sheet or layer of cloud belongs to the upper or the lower system. In such cases the observer will be greatly assisted by remembering how the clouds have assumed the shapes he sees, whether by the gradual subsidence of the higher forms, or by the ascent of the lower clouds.

We now come to the lower clouds.

Stratus is a form of cloud about which there has been a persistent misconception, as there has been a tendency to give that name to every thin layer of cloud seen on the horizon. Howard's definition of stratus is 'a widely extended continuous horizontal sheet increasing from below upward.' It is a sheet or layer of cloud, of uniform thickness generally. It has but little variety of light and shade, and belongs essentially to the lower regions of the atmosphere, so much so, that Howard speaks of it as 'ground fog,' the cloudy formation which spreads over low grounds in the evening, and disappears as soon as the temperature rises in the morning.

Stratus is generally a fine-weather cloud, appearing during the evenings and mornings of the brightest days. At times it overspreads the whole sky in the form of a low gloomy foggy canopy, the atmosphere being more or less foggy under it. All low detached clouds which look like a piece of lifted fog, and are not in any way consolidated into a definite form, are *stratus*.

Cumulus is a very common form of cloud and frequently appears in summer. It has a rounded, or conical, shape, and often springs from a horizontal base. Hence its name of 'wool-pack cloud.' It is produced by an ascending current of air, of which the vapour is rapidly condensed, and has a tendency to collect in rudely globular masses, such as are formed by the steam escaping from a locomotive on a frosty day.

Cumulo-stratus is defined by Howard to be 'the cirro-stratus blended with the cumulus, and either appearing intermixed with the heaps of the latter, or *superadding a widespread structure to its base.*'

This is also a common cloud form. It is the *cumulus* as it were changing into a *nimbus*. It is dark and flat at its base, and is traversed by horizontal lines of dark cloud. It not unfrequently covers the whole sky, and thus is one of the types which go to form the *pallium* of Poëy.

It need scarcely be stated that all these appearances of horizontal bases to clouds indicate the presence of strata of air of different temperatures. If the lower stratum be at a higher temperature than the upper, it can contain more moisture, and accordingly can dissolve any portions of the cloud which descend into it, so that the cumulus cloud will appear to stand on the plane of separation of the two layers.

In some cases the phenomenon is noticed of a cumulus cloud apparently reversed, a globular formation being developed on its under, instead of its upper surface. This cloud is called in the Orkneys the 'pocky cloud,' and is a well-known sign of storm. It is also not very uncommon at the time of thunderstorms. In the case of the formation of this cloud it is probable that a warmer and damper stratum is superposed on a cold one, so that there is a condition of unstable equilibrium, likely to produce a storm.

Nimbus, also called *cumulo-cirro-stratus*. This is the rain-cloud, defined as 'a cloud or system of clouds from which rain is falling.' Whilst on the horizon or as it advances towards the observer, its front frequently presents a marked outline, like that of a very heavy *cumulo-stratus* with rain falling from it and with some *cirrus* above. When it has covered the whole sky it is so concealed by falling rain that it generally assumes a uniform dark appearance.

It is, however, evident that as rain may fall from

clouds of various types, the description of *nimbus* as a compound cloud leaves very much to be desired.

Equal in importance to observations of the *form* of the clouds are observations of their *motion*. The clouds of the lower stratum usually move in the same direction as the wind felt on the surface of the earth, at least in flat countries. Among mountains, of course, the surface wind is affected by local eddies. At all times, however, even the lowest clouds move far more rapidly than the wind close to the ground.

As regards upper clouds the case is widely different, for their motion almost invariably makes an angle with that of the wind below, and by the study of this motion, as has already been said, much light has been thrown on the laws which govern atmospherical currents.

It is also important to observe the *amount of cloud* visible in the sky at any time. This is estimated according to a scale 0–10, 0 being a clear blue sky, 10 a sky entirely overcast. Such an estimation is more or less unsatisfactory, as it is entirely relative to the position of the observer, for a layer of stratus, which, if it were in the zenith, would cover a considerable proportion of the sky, only presents to the eye a thin section when it is seen on the horizon.

Fog, of course, gives the appearance of an overcast sky (10 of the scale), but it is only to be entered as 10 when the observer is entirely enveloped in it and cannot see any blue sky.

The amount of cloud has an annual as well as a daily period. The former is not as perceptible in this country as in continental climates. The latter appear in a very marked way in tropical countries, where it connection with the ascending currents of air is unmis takeable. In some parts of Brazil the clouds at certai seasons collect every afternoon to form a nimbu: resulting in a thunder-storm, and the same phenomeno is said to occur frequently at the Lake of Como.

CHAPTER VIII.

RAIN, SNOW, AND HAIL.

Rain.—We now come to the most important mode of restoration to the earth of the moisture which is abstracted from its surface by evaporation, and which returns to it most frequently in the liquid, but occasionally in the solid, state.

It is comparatively an easy matter to collect rain, but the case is different as regards hail and snow. Indeed the drifting of snow produces such irregularities in the distribution of the fall that any estimation of the general amount collected over a given area must be made more or less at random.

The simplest form of rain-gauge is Howard's consisting of a funnel to catch the rain and a Winchester quart bottle to receive and collect it. The only objection to this is the liability of the bottle to break in frost or with hard usage. The water collected is always measured in a graduated glass, which shows true inches for the aperture of the gauge. Glaisher's gauge is a modification of Howard's, with a copper can, instead of a bottle, to collect the rain as it flows from the funnel, the delivery-tube being, moreover, bent in a U form, in order to check evaporation from the receiver below. The chief improvements recently effected in the construction of these gauges are the straightening of the small tube just mentioned, to prevent its being choked with dirt, and the addition of a vertical cylinder on the top to retain snow when it falls, and prevent its being blown out of the gauge, as would be very likely to happen if the funnel were shallow. This improvement was devised by the Meteorological Office in 1868. Such gauges are now known as 'Snowdon gauges.' Fig. 23 shows this gauge in its most recent form. The diameter of the cylinder is 8 inches.

FIG. 23.

Rain Gauge with Measuring Glass.

In addition to the simple gauges just mentioned there are endless other patterns. We have float-gauges adapted for use at mountain stations which are rarely accessible; various forms of automatic gauges which are moved either by the wind or by a clock, so as to separate the amount collected either according to the direction of the wind or to the hour at which the fall took place; and finally, we have gauges, like Beckley's, which show continuously the rate at which the rain comes down.

The measurement of rain seems an easy matter, but practically it requires considerable precautions.

Firstly we infer the amount collected over a large district from that caught in a small funnel, so that great care is requisite to ensure that we know exactly the sectional area of the funnel at its top. Mr. Symons and others have recommended the use of 5-inch gauges. The Meteorological Office, however, prefers the 8-inch gauge—the size originally adopted by Mr. Glaisher—as affording greater security against error in measurement, the amount to be measured being larger than if the diameter were only 5 inches.

If we know the size of our gauge we must ensure that this size will not be changed, as would be the case if the shape of the aperture were altered. Most gauges are circular, and, as a circle encloses the greatest area of all closed curves of the same perimeter, the least dint, or squeezing in, of the sides of the funnel must make it indicate too small an amount of rain.

Secondly, precautions must be taken that there shall be no loss, by evaporation, of the water collected in the gauge. This is very likely to occur with passing showers on a hot day, and Mr. Glaisher attempted to meet the difficulty by curving the delivery-tube.

Thirdly, the greatest care is required as to the exposure of the gauge, both in regard of its being sheltered by trees or walls and in regard of its height above ground. It is unnecessary to insist on the importance of a perfectly open site. Every one takes advantage of a tree or wall for shelter from a shower. Such obstacles, even at some distance, produce irregular currents of air and disturb the equable fall of the rain. No object ought to subtend a greater angle from the horizon than 20° in any direction from the gauge. The gauge must be firmly fixed, so that it shall not blow over, and its rim must be perfectly level.

The height above the ground is a matter of no less importance in rain measurements. It has been generally decided in this country to adopt one foot above the ground for the height of the gauge, but this rule has not been followed abroad. It has long since been established that more rain is collected at the surface of the ground than at any elevation above it. The reason of this is now clearly understood, owing to the researches of Mr. Symons and Mr. Dines.[1] The difference is simply due to the mechanical effect of a building or other support, on which the gauge is placed, in producing eddies, and it increases with the force of the wind.

In connection with the placing of gauges on roofs or at a height above the ground, it is of importance to draw attention to the great influence exerted by a vertical wall on the indications of a gauge placed upon it. If you are standing on the edge of a cliff against which a violent wind is blowing, you find yourself in a calm, owing to the upward blast produced by the impact of the wind on the cliff face, and interfering with the current, which would otherwise sweep over the top of the cliff. It is a well-known fact that sheep will crowd to the edge of a 'weather' cliff in a gale, as they know that a few yards back from the edge they will feel the wind with increased force. The rain drops will accordingly be carried over and past a gauge on a wall when the wind is blowing against that wall.

The late Rev. Dr. Robinson, of Armagh, has told the writer that he once had a very perfect illustration of this fact. He had constructed a peculiar gauge, consisting of a *sphere*, erected over a funnel, which, therefore, exposed the same section to the rain at whatever angle it might be driving. Under ordinary circumstances this gauge collected more than one of the ordinary pattern. It was, however, erected on the parapet of the flat roof of the Observatory. The distance from the dormer window, which gave access to

[1] *Report of the British Association*, 1881, p. 551. *British Rainfall*, 1880, p. 13; 1881, p. 41.

the roof, to the gauge was about 30 feet. During one heavy gale Dr. Robinson went up on the roof to see how the gauge was acting. It was raining so heavily that he was wet through while walking over the slight distance of 10 yards, and yet not a drop of water was in the gauge; all had been blown over it by the eddies.

The British Rainfall System, founded by Mr. G. J. Symons, F.R.S., in 1860, and carried on under the auspices of the British Association until a recent period, has now grown into an extensive organisation numbering over 2,000 observers. This success is almost entirely due to the energy of its founder, and the result is that we possess more accurate statistics of the rainfall of the United Kingdom than of any other country in the world.

Still, however we may multiply rain observers so as to secure a complete record of the rainfall, we are a long way from a proper knowledge of the true method of protecting ourselves against the rain when it falls, and of dealing with it when it has fallen.

A gallon of water weighs 10 lbs., and, if spread out in a layer an inch thick will only cover an area of two square feet. An inch of rain gives 100 tons of water per acre, or 60,000 tons per square mile. Now, on July 14, 1875, more than 5 inches of rain fell in Monmouthshire, and, accordingly, there were 500 tons of water to be collected and stored for every acre. Such a fall is, fortunately, extremely rare in the British Isles, and, if we were to provide reservoir space to store it, we may be certain that the utilitarian tendencies of the residents in the neighbourhood of the dam would lead them to use the basin for agricultural purposes, and that we should have a prodigious outcry for compensation if the reservoir were ever filled and their precious pasture submerged.

This is, however, one of the grand engineering problems of the future, in the face of an increasing price of fuel, being one mode of utilisation of the natural sources of power which are lying within our grasp. We hear about concentrating and storing up solar heat; but here is an agency more directly manageable, and one which is being seriously modified by civilisation.

Owing to the clearing of forests and the draining of land, rivers are becoming less navigable in their upper waters, while the floods are becoming more serious on the lower parts of their courses.

These apparently contradictory statements may be reconciled as follows. The first result of the removal of the growing timber is the disappearance of the roots of the trees, which formed a network to hold the soil together. The rain falling on the bare hill-side sweeps off the earth with the loose gravel and boulders, and deposits it on the plains below, filling up the river beds, or at least producing serious shoals. The same action causes the rain-water to run off the land more rapidly than would be the case in the natural state of things, and consequently the water, instead of being retained in the earth, comes down at once and produces a flood in the lower parts of the river.

The attention of Continental Governments has been seriously attracted by the importance of this question, and the sooner we in this country look to it, not only for India, but also for the British Isles, the better for us.

The amount of rain which falls over the globe will be discussed in a subsequent chapter; but I may say here that in these islands the annual quantity at sea-level ranges from 60 or 80 inches, on the west coasts of Ireland and Scotland, to about 20 inches on the east coast of England. In some localities, however, the fall is much greater, amounting to 154 inches, on the average of six years, at Seathwaite, in Borrowdale, at the height of 422 feet above the sea.

The quantities which fall in particular showers are often very great, and though we in the temperate zone cannot rival tropical experiences, we have something to boast of. Arago reports that, on October 9, 1827, 30·9 English inches of rain fell in 22 hours at Joyeuse. At Gibraltar, October 25, 1836, when a waterspout broke over the town, 30·11 inches were collected.

About London a fall exceeding an inch in twenty-four hours is comparatively rare (although, August 1, 1846, 3·12 inches were collected in St. Paul's Churchyard[1] in 2 hours and 17 minutes), but on our west coasts this amount is often exceeded. October 24, 1849, 4·37 inches were collected at Wastdale Head; June 30, 1881, 4·80 inches at Seathwaite; April 13, 1878, 4·6 inches fell at Haverstock Hill, in London, and the fall of 5·36 inches in Monmouthshire, July 14, 1875, has been already noticed. Such accounts may be indefinitely multiplied.

Let us now examine in more detail the causes of rain. It is produced by the chilling of air more or less charged with moisture. This may be effected in various ways, of which the following are the principal:—

1. The ascent of a current of damp air into the colder regions of the atmosphere.

2. The contact of warm and damp air with the

[1] *Report of the British Association for* 1846, Part II., p. 17.

colder surface of the ground, as in the case of our own west coasts in winter, where the land is colder than the sea surface.

3. The mixture of masses of hot and cold air.

The first of these is the most efficacious, and the process is almost always in action in the torrid zone. Dr. Hann has calculated that the temperature of ascending air will be lowered 1° F. for 182 feet (or 1° C. for every 100 metres), *supposing that no moisture is condensed*. We see, therefore, that the change of temperature induced in perfectly dry air by rising above the tops of the Andes—say to a height of 25,000 feet—would be very great, amounting to nearly 140° F. In reality, however, the ascending air is never perfectly dry, and as vapour is extensively condensed in the course of the ascent, the latent heat set free by that process will retard the rate of cooling of the air.

Near the equator, we may say, speaking in general terms, there is all but constant precipitation in heavy showers, and the fall is greatest when the weather is hottest. Dampier gives a very lively picture of the rain at Gorgonia, off the coast of Panama, when he says that, when he and his mates were drinking chocolate in the open air, it rained so heavily that some of the men declared that they could not empty their calabashes, for they could not drink up the water as fast as it fell into them.

The second mode of production of rain is that to which most of the rain which falls on our own west coasts is due. If we look at any rain map of the globe we shall see that all the continents, in the temperate zone, have wet coasts to the westward, but their west coasts are also far more mountainous than their east coasts.

Now, although the land is colder than the sea-surface, it is only the very lowest stratum of air which can be chilled by actual contact with the ground. It is undeniable that the westerly winds bring moisture to the coast, a portion of which is deposited in the form of rain as they pass inland; but, as the coast is mountainous, and the air is forced to ascend, the previously-described mode of condensation comes into play, producing the drenching rains so common on the seaward slopes of our western hills. In this way the climate of our western shores in winter is rendered very mild, owing to the abundance of latent heat set free by the continual condensation of vapour.

The last mode of effecting condensation is one which was formerly held to be of great importance, the mixture of masses of air of different temperatures. It is, however, not very easy to see how this mixture could be brought about, and, even if it were, as Dr. Hann has shown,[1] the result would not be great.

'If a mass of saturated air at 77° F. be mixed with another at 32° F., as the weights of vapour per cubic foot are 10·0 grains and 2·1 grains in the two masses respectively, the arithmetical mean temperature of the mixture (if the air be supposed dry) would be 54°·5, and the arithmetical mean of the weight of vapour 6·1 grains. But at 54°·5 a cubic foot of air can only contain 4·8 grains, so that from every cubic foot 1·3 grains would be condensed by a fall to that temperature. Now the condensation of 1 grain of aqueous vapour sets free 0·093 units of heat, which are available to warm up

[1] *Allgemeine Erdkunde*: Hann, Hochstette and Pokorny. (3rd ed.). Tempsky, Prague, 1881, p. 105.

the air,[1] so that the temperature cannot fall as low as we have supposed.

'In fact calculation shows that the temperature of the mixture would be $58°·7$ (instead of $54°·5$), and the weight of vapour condensed would be only 0·6 grains per cubic foot. The amount of rain such a condensation would produce would be quite insignificant; a column of air 1,000 feet in height yielding only about 600 grains, or one-twelfth of a pound of water, per square foot of section. This would correspond to a rainfall of ·048 inch—not quite half a tenth of an inch.

It is evident, therefore, that the mixture of volumes of air cannot be very effective in the way of bringing about precipitation, and the case which has been assumed is very extreme, and could hardly occur in nature.'

I have said a good deal about the effect of rainfall in warming the air; the greatness of this effect may be seen from the following quotation from Haughton, 'Physical Geography,' p. 126 :—

'*One gallon of rainfall gives out latent heat sufficient to melt 75 lbs. of ice, or to melt 45 lbs. of cast iron.*'[2]

[1] 1 grain of aqueous vapour on condensing gives out heat enough to raise 536 grains of water 1° C. Now, the unit of heat is 5,760 grains of water raised through 1° C., so 1 grain of vapour, when condensed, sets free $\frac{536}{5760} = 0·093$ units of heat. Now, the specific heat of air is 0·2375, and the weight of a cubic foot of air is 560 grains, so the condensation of 1 grain of aqueous vapour raises the temperature of 1 cubic foot of air $\frac{5760 \times ·093 \times 9}{560 \times ·2375 \times 5} = 7°·25$ F.

[2] The following table shows the exact amount of latent heat contained in water at various temperatures :—

Temperature	Pounds of Ice melted	Temperature	Pounds of Ice melted
40° F.	76·15	60° F.	75·16
45° F.	75·91	65° F.	74·94
50° F.	75·65	70° F.	74·69
55° F.	75·43	75° F.	74·44

'From this datum it is easy to see that every inch of rainfall is capable of melting a layer of ice upwards of 8 inches in thickness (exactly 8·1698 inches) spread over the ground.'

Dr. Haughton goes on to calculate that, on the west coast of Ireland, the heat derived from the rainfall is equivalent to half that derived from the sun.

Snow.—We all know that, if water be cooled sufficiently, it will pass into the solid condition. When, therefore, the aqueous vapour in the air is condensed at a very low temperature it freezes at once, and falls as snow or hail.

There is a story of a ball at St. Petersburg at which the atmosphere became so oppressive that ladies began to faint. Gentlemen climbed up and broke the windows to let in air, and the company were disagreeably surprised by finding that snow began to fall in the room. The moisture in the air within the building passed at once into the solid form, under the influence of the cold air from outside.

At all places for which the mean temperature is below 32°, the deposit of moisture, speaking in general terms, will ordinarily take place as snow. Snow also often falls when the temperature is above the freezing-point, but it is never observed when the temperature is much above that point, and, accordingly, there are many parts of the world where it never falls at sea-level.

The mode of measurement of snow is to thaw the quantity collected in the gauge and read off the amount in the measuring-glass. If the fall is very large, the easiest way of melting the snow is to add a measured quantity of warm water, which is of course to be subtracted from the final reading. Other suggestions as to snow measurement will be found in the 'Instructions' already mentioned.

Snow is much less dense than rain, and—*on a very rough estimate*—a foot of snow yields about an inch of rain; so, when we hear people saying that 6 feet of snow have fallen at one time, we may be sure that there is some exaggeration. A uniform fall of a foot at one time is very unusual in these islands, but as all the snow which falls does not thaw at once, the amount on the ground is often increased by successive falls. When the wind drifts it, the snow in sheltered spots may, and does, attain a depth of several feet.

The forms of snow-crystals are often of extreme beauty. These forms belong to what is called the hexagonal system—the same as that to which quartz

FIG. 24.

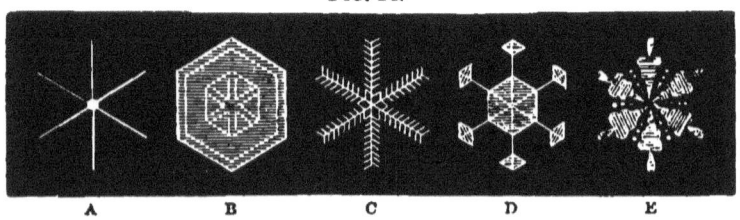

A B C D E

Types of Snow Crystals. From a collection of 151 specimens drawn by J. Glaisher, F.R.S. (5th Report of the Council of the British Met. Soc., 1855.)

and calc-spar belong. The crystals are either hexagonal plates or six-pointed stars, of great variety as to their character, but so formed that the angles shall bear a close relation to those of a regular hexagon, *i.e.*, shall be multiples of 15° or 30°. The above are a few specimens.

The large flakes of snow which we see, especially during warmish weather, are formed by the agglomeration of smaller flakes, and this process can only take place when the air is damp. In very dry weather the snow falls in the condition of small pellets, like 'soft hail.'

Sleet is a mixture of snow and rain.

Hail is much denser than snow, being more like ice. There are two types of it—soft hail and true hail.

Soft hail (*Grésil*, Fr.; *Graupel*, Germ.) falls chiefly in winter and spring, when the air is dry: its grains are not very hard, and are simply rounded pellets.

True hail is a very different thing. It is usually composed of concentric layers of hard and soft ice, and

FIG. 25.   FIG. 25A.

 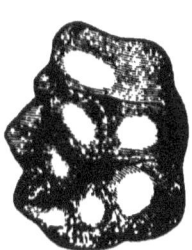

the stones assume an irregular shape. Sometimes they have a rough exterior, owing to the formation of crystals, which are generally found to have lost the sharpness of their edges either in their fall through the warm air or else on the ground before they were picked up.

The woodcuts show specimens of the various structures of hailstones, several of which represent stones which fell near Tiflis, in Georgia, in May and June 1869, and were picked up and drawn by Professor Abich, a well-known mineralogist.[1] Fig. 25 was the commonest form of occurrence. It consisted of a spheroidal agglomeration of small stones frozen together,

[1] *Ueber Krystallinischen Hagel im unteren Kaukasus.* Von H. Abich. Hölder, Vienna, 1879.

the mass was usually half thawed before it was picked up and drawn. Fig. 25A (p. 143) shows the same structure more clearly.

Fig. 25B represents a very remarkable type, of which several specimens occurred, and of which the crystalline form had not disappeared before the stone was drawn.

The crystals with which the nucleus was studded exhibit forms closely resembling those of calc-spar and specular-iron. The diameter of these stones was about 2 inches.

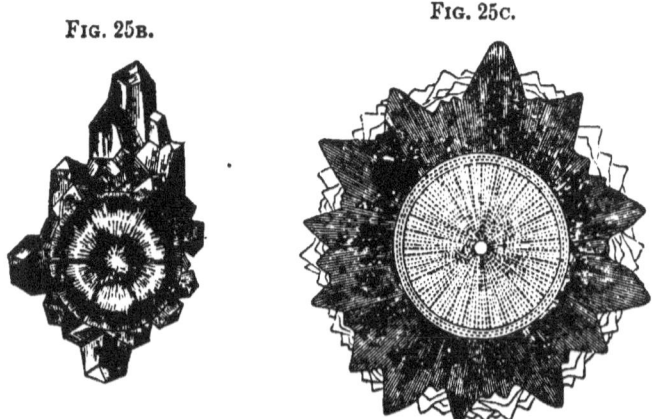

FIG. 25B.   FIG. 25C.

Finally 25c is an attempt at a representation of the section of a stone figured by Delcros, which fell during a tremendous storm on the night of July 3—4, 1819, at La Braconnière, department of Mayenne, in France, and was described by Arago and Delcros.[1] These stones consisted of a small spherical kernel, surrounded by a shell of dense bluish ice, less perfectly transparent, and showing traces of a concentric arrangement, and of a radial structure from the centre to the circumference. The whole was coated externally with pyramidal crystals.

[1] *Poggendorff's Annalen*, vols. lxviii. and lxxxix.

Very large hailstones, weighing several ounces, have been sometimes observed. Some have been picked up in the Orkneys as large as goose eggs, but each of these very large stones is a mass of smaller ones which have come together during their fall.

Stories of hailstones weighing pounds, or even as large as an elephant, like that reported to have come down near Seringapatam, in Tippoo Sahib's reign, are, of course, ludicrous. The germ of truth there is in them arises from the possibility of a number of stones falling successively into a hole like a well, and freezing into a single mass.

The formation of hailstones, however, whatever be their size, is very difficult to understand. No sufficient explanation has yet been offered of their structure, or of the agency which enables them to remain suspended in the atmosphere before they fall.

In the explanation already given of glazed frost, the occasional presence in the atmosphere of water cooled below its freezing-point has been mentioned. Liquids in such a condition solidify instantly on agitation or contact with a foreign body. We may, therefore, suppose the origin of the initial hail particle to be a drop of water, in a state of superfusion, caused to congeal by some sudden impulse of wind, which produces agitation. A number of such drops might easily congeal together into a single mass, but it is hardly conceivable that structures of such complexity as some of those shown above could be the product of any instantaneous action, and therefore these must have been suspended in the atmosphere for some time.

The oldest theory of the formation of hail is that of Volta, who supposed that, as hail is generally associated with thunder and lightning, the hail-pellets were in a state of constant oscillation between two oppositely electrified clouds. Condensation, according to him, was always going on, and the stones grew in size until they became so heavy that they broke through the lower stratum and fell to the earth.

Dove broached a different theory, which is that hailstorms are always whirlwinds, but with their axis almost horizontal instead of vertical. The air then sweeps the growing hailstone round and round from hot to cold stratum alternately; water settles on it in the hot, and is frozen in the cold, layer. Ultimately, either by its own weight or by centrifugal force, it comes outside the influence of the rotatory motion of the whirl and falls to the ground.

In these islands we, fortunately, know but little of destructive hailstorms. On the Continent we are almost everywhere met by the frequent advertisements of Hail Insurance Offices, which are comparatively rare here. This immunity of ours is due to our insular climate, and its consequent freedom from extremes of temperature.

The geographical repartition of rainfall, and its seasonal distribution in different latitudes, will be considered later.

The law of the fall in the daily period has not been very much studied, as automatic gauges are comparatively modern inventions, and the main fact which is known on the subject is that as, between the tropics, the wettest portion of the year is the hottest, so there also the early afternoon, the hottest part of the day, is the wettest.

Caldcleugh[1] says of Rio Janeiro that it used to be the fashion to state in invitations for the afternoon whether the guests were to assemble before or after the thunder-storm, which came on regularly at a certain hour. In fact, in the tropical regions the clouds frequently form in the forenoon, the rain falls in the afternoon, and all the night through the sky is cloudless.

In a recent paper on the rainfall of Cherrapunji, by Professor John Eliot, read before the Meteorological Society, and printed in their 'Quarterly Journal' (vol. viii., p. 41), it is stated that, during the wet season there, about twice as much rain falls by night as by day.

In Europe the distribution of the fall has not been studied for long or for many stations. Prague is that which affords the longest series of hourly records, but material also exists, and has been discussed, for Greenwich, Vienna, Modena, St. Petersburg, Zechen, Berne, and Coimbra, as well as for some other observatories. The result shows that the law is a very complicated one. Almost all these stations exhibit three maxima and three minima in the twenty-four hours, and at most of them the absolute maximum, the wettest portion of the day, occurs at two or three o'clock in the afternoon. The figures for Berne, however, form an exception, for there the afternoon maximum entirely disappears, and is replaced by one, very strongly marked, at 10 P.M. The subject demands further study.

[1] Daniell, *Meteorological Essays and Observations*, 1st edition, 1823, p. 335.

CHAPTER IX.

THE WIND.

WE have already discussed the conditions of the atmosphere as regards its temperature and its pressure, and we have now to consider what the immediate effects o these conditions are in disturbing its equilibrium and determining its motion.

Air in sensible motion is wind, and wind is produced by differences of pressure, which are themselves ultimately, and in the main, attributable to differences of temperature.

We must first say something about the way in which the direction and force of wind are estimated.

The direction of the wind is recorded according to the points of the compass, and for ordinary observations the estimation according to the eight principal points (N., N.E., E., S.E., &c.) is close enough Care must always be taken to enter the wind to *true* bearings, for as in these islands the north end of the compass needle lies to the west of the geographical north, it is evident that all 'compass' bearings will be affected by this *variation*, as it is called, of the magnetic needle, which practically amounts to two points. The variation changes slowly from year to year; it is at present diminishing in the British Isles, and its value varies from about 19° in the south-east of England to about 25° in the north-west of Ireland. The lines of equal variation run N.N.E. and S.S.W. The mode of recording the direction numerically is given in the note.[1]

The instrument employed for giving the direction of the wind is the ordinary wind-vane. In using this apparatus care should be taken that the vane is long enough to be easily affected by the wind, that the bearings are kept well oiled, and that the cross, indicating the cardinal points, is set by the true meridian.

[1] The compass card is divided into 32 parts called points. The number of degrees in the entire circumference being 360, the points in their numerical order, with their angular difference from North, are given in the following diagram:—

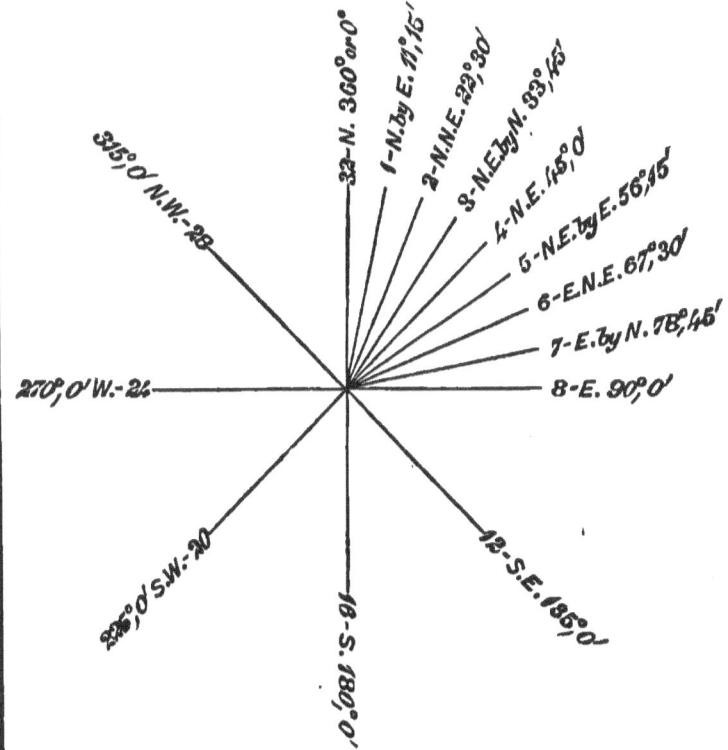

From this it will be seen that a difference of a point, in the bearing of an object, amounts to 11° 15'.

As regards the motion of the air we may measure either its pressure or its velocity.

The former is by far the earliest method of registering the force of wind, and the oldest form of instrument is described, apparently by Robert Hooke, more than two hundred years ago, in the 'Philosophical Transactions,' vol. ii. p. 444, an arrangement which has lately been recommended for universal adoption by Prof. Wild, of St. Petersburg. It consists of a rectangular plate of sheet iron, suspended from its upper edge, like the sign-board of a public-house. It is evident that such a plate, when the wind blows against it, will be displaced from the vertical position, and that we can form some judgment as to the force of the wind from the angle at which the plate rests, but the instrument is somewhat imperfect.

Another form of pressure gauge is that of Lind, which consists of an inverted siphon tube half filled with water and graduated on the closed arm (fig. 26). A little water is introduced into the tube, sufficient to fill both legs up to the point marked 0. The force of the wind is measured by the height of the column of water which the wind is able to maintain in the longer arm above the level of the water in the shorter. The instrument is held in the hand, and the opening of the leg of the siphon, which is bent at a right angle, is turned to the wind. Such an experiment is necessarily rough.

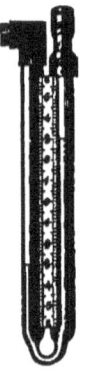

FIG. 26.

Lind's Anemometer.

The most complete forms of pressure anemometer are those of Osler and of Cator, in which the pressure is received by a plate, the motion of the plate being communicated to a pencil which marks the force on paper. In Osler's instrument the resistance is furnished by springs, in Cator's by a series of levers.

It has been objected to Osler's instrument, that springs, from their continual exposure to weather, cannot always preserve the same strength and elasticity, and that the results must be affected by any change in these conditions.

Cator's instrument is described in the 'Instructions,' already so often referred to.

The main difficulty with regard to both instruments consists in the fact that neither the results obtained from different instruments with the same size of plate, nor those from the same instrument with pressure plates of different sizes, show such an accordance *inter se* as might be wished; so that we cannot say what is the meaning of the indications of the instrument, or what is the best size to adopt for the normal plate. That usually employed is one square foot.

FIG. 27.

Robinson's Anemometer.

Notwithstanding this objection, pressure anemometers are of great importance as being the only instruments at present which can give us any record of individual gusts of wind or of any sudden changes in wind force.

Of velocity anemometers the only form which is now practically in use is that of the late Dr. Robinson (fig. 27). In this instrument the velocity is measured by the rate of revolution, in a horizontal plane, about a vertical axis, of four hemispherical cups fixed to the extremities of the arms of a cross. The motion of the axis is transferred by gearing to the recording portion of the apparatus, which may be either a system of dials, to be read off at definite intervals, as shown in the figure, or an arrangement for marking each mile of wind on a strip of paper moved by clockwork; or, finally, as in the full-sized anemographs, a pair of brass tracers which mark the velocity continually on metallic paper, stretched on a drum which is caused to revolve at a regular rate. In these latter instruments the direction is recorded by the tracers in a similar manner to the velocity, the moving power which gives the direction being a double windmill vane, on the plan devised by R. Beckley.

According to the theory originally proposed by Dr. Robinson for his instrument ('Transactions of the Royal Irish Academy,' vol. xxi. 1850), the velocity of rotation of the cups was taken to be independent of the length of the arms and the size of the cups, and it was assumed that they always moved with one-third of the velocity of the wind, so that the factor 3 was employed for deducing that velocity from the indications of the anemometer. Practically, however, it has been found that these two assumptions are not quite correct, and Dr. Robinson himself, after a long and careful series of experiments, has published his final results,[1] in which he admits that the factor varies with the pattern of instrument used, and that it is less than 3.

The most concise statement of the present position of the question is given by Prof. Stokes in a paper in the 'Proceedings of the Royal Society,' vol. xxxii. p. 170.

---

[1] *Phil. Trans.*, vol. clxix., p. 777; vol. clxxi., p. 1055.

He finds (A). 'That the factor varies materially with the pattern of the anemometer. Among those tried, the anemometers with the larger cups registered the most wind, or, in other words, required the lowest factors to give a correct result.

(B). 'That with cups of 9 inches diameter, on arms 24 inches in length (the pattern used by the Meteorological Office), the register gives about 120 per cent. of the truth, requiring a factor of about 2·5 instead of 3. Even 2·5 is probably a little too high, as friction would be introduced by the centrifugal force, beyond what occurs in the normal use of the instrument.

(C). 'That the factor is probably higher for moderate than for high winds; but whether this is solely due to friction, the experiments do not allow us to decide.'

In fact, the smaller instruments commonly in use yield results 15 or 20 per cent. below the larger ones, but the subject urgently requires further investigation.

It is also evident that the action of wear and tear, if the instrument be not kept constantly oiled, must have a very material influence on its indications.

There is this inherent difficulty about all anemometrical measurements, whether of pressure or of velocity, that they are seriously affected by the position of the instrument. If it be erected on, or even near, a large building, or if there be high trees in the vicinity, it will indicate much less wind than if it were in a perfectly open country; and the results of recent experiments have thrown doubt on the possibility of comparing, with any degree of minuteness, anemometrical data from different stations.

This fact alone is sufficient to condemn most of the determinations of absolute velocities made hitherto with small anemometers, even though the instrument may have been quite correct.

Various instruments have been devised, employing methods of electrical registration for the velocity, and allowing of the cups being erected at a distance from any building, but it seems all but impossible to erect the instrument so that it shall be removed out of

FIG. 28 A.

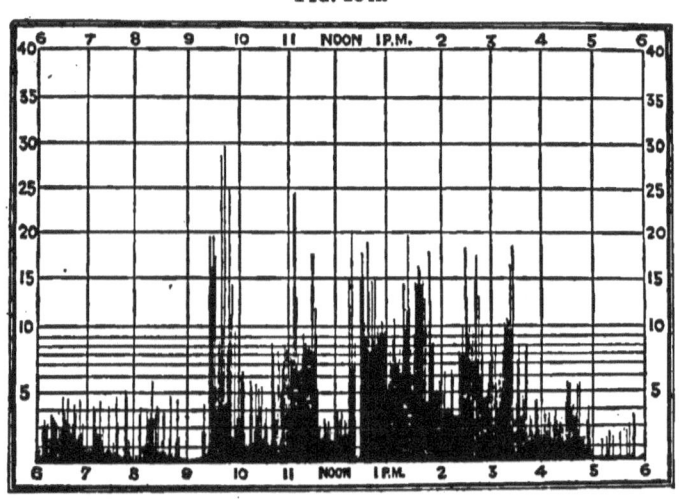

Records of the Osler's Anemometer at Glasgow Observatory, October 14, 1881.

the sphere of the disturbing action of the irregular surface of the ground on the current of air passing over it. All that can be said at present is that the results from each station must be compared *inter se*.

The best table of the relations between pressure and velocity is that given by Sir H. James,[1] which is computed from the formula that $P = v^2 \times \cdot 005$, and

---

[1] *Instructions for taking Meteorological Observations with Tables*, &c., by Colonel Sir H. James, R.E., 1861, p. 32. The pressure varies as the square of the velocity. The square of the velocity in miles per hour multiplied by 0·005 gives the pressure in lbs. per square foot, or $v^2 \times \cdot 005 = P$.

extends up to the pressure of 50 lbs. per square foot, and the velocity of 100 miles an hour. Any one, however, who has ever compared the simultaneous indications of an Osler's and a Robinson's anemometer, placed side by side (figs. 28 A, 28 B), will at once admit the great difficulty of converting the one into the other. It will be seen that the pressure traces (in pounds per square foot) vary in length from instant to instant, while the velocity lines (representing miles run in the hour), at the contracted time-scale we are obliged to use, may show but little trace of any inequality of speed.

The pressure in storms is evidently exerted in distinct and separate gusts, each affecting a limited superficial area; for it is inconceivable that if the pressures which are sometimes recorded by our instruments were felt over the surface of a wall or chimney the structures could stand. It is a well-known fact that storms occasionally cut lanes through thick plantations, leaving the trees standing on either side. This was particularly noticed at Alnwick in the storm of October 14, 1881, and FitzRoy ('Weather Book,' p. 265) notices it as a common feature of hurricanes.

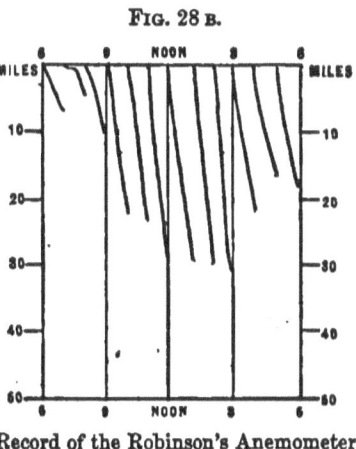

FIG. 28 B.

Record of the Robinson's Anemometer at Glasgow Observatory, October 14, 1881.

The author has himself been a witness of an accident which illustrates the way in which damage may possibly be done in storms to individual trees or buildings, while objects adjoining them escape comparatively uninjured. In the summer of 1867, a gas explosion took place in a house in London. The gas caught fire in the dining-room, of which the window was blown out; a second or so later another explosion took place in the hall, which had a folding inner door besides the hall-door. One side of the inner door was torn off its hinges, the other untouched. The hall-door was broken and blown across the street, while the panes of glass beside it were not even cracked, and the large window on the stairs facing the hall-door escaped without injury.

Here, then, we have sufficient motion produced in a mass of air to tear a heavy door off its hinges, while at the same time the circumjacent air was comparatively undisturbed, and the panes of glass were left whole. We cannot, of course, compare, except in a very distant manner, the behaviour of the wind in a storm to that of the air in a gas explosion, but it is evident that extraordinary records, such as those of a pressure of over 70 lbs. per square foot at Bidston Observatory, February 1, 1871, *may* have been a correct indication of the force exerted on the plate, but not of the violence of the wind at the distance of even a few feet from it.

This statement derives additional confirmation from a remark by Mr. Blanford in a paper on the climate of Bengal. ('Proceedings of the Asiatic Society of Bengal,' Nov. 1873.) He says:—

'The highest (pressure) that has been registered in Calcutta by an Osler's anemometer is 50 lbs. to the square foot, but this was in a storm of no remarkable violence and one which did but little injury in Calcutta.

'In the far more severe storms of November 2, 1867,

and October 5, 1864, the anemometer was blown away under a pressure of 36 lbs. to the square foot, so that no register of their maximum force was obtained.'

It must, however, be stated that there is no evidence in existence as to the area over which individual gusts may extend, and that although in some cases this may be very limited, in others it may be sufficient to cause serious damage over a large surface. It would seem advisable for engineers in all structures to make allowance for a possible pressure of at least 40 lbs. per square foot.

To return to the subject of the relation of velocity to pressure or force, either measured or estimated, the most complete table of the older results is one given in the 'Edinburgh Encyclopædia,' and there cannot be a better illustration of the hopeless state of confusion into which the subject has been brought. The following are a few instances of pressures, velocities, and descriptions given in the table, with the respective authorities:—

Pressure; lbs. per square foot	Velocity; miles per hour	Description	Authority
9·963	49·69	Great storm	Denham.
21·485	74·69	Great storm	La Condamina.
46·875	107·80	Most violent hurricane	Lind.
49·200	110·48	Hurricane that tears up trees and throws down buildings	Rouse.
58·450	120·37	Observed by	Rochon.

We are not informed of the effect of the wind which Rochon observed; if it did more than throw down buildings it must have been hard to register its force.

In the 'Weather Book,' Admiral FitzRoy gives a table, contained in a letter from Mr. Glaisher (dated in

1858), in which the several degrees of the so-called 'land' scale (0 to 6) and their sub-divisions are represented by pressures per square foot varying from 1 oz. up to 36 lbs. This, however, only shows that the land scale, as understood at the time the letter was written, was insufficient to represent the extreme forces of wind which may possibly occur; for, not to speak of the record at Bidston above referred to, we find that pressures up to 40 lbs. have not unfrequently been registered, *e.g.* 53 lbs. at Greenwich, October 14, 1881; 42 lbs. at Glasgow, January 24, 1868. These, therefore, would correspond to forces above the highest figure of the scale.

The oldest method of observation of wind is by sensation or by estimation. This is necessarily somewhat rough, as it must depend to a very great extent on the individual observer. Nevertheless, it is the only mode of obtaining wind observations at the majority of stations, and with experience can yield useful results.

Sir F. Beaufort, when in command of H.M.S. 'Woolwich,' in 1805, devised a scale for these estimations, having relation to the pressure of the wind on the sails of a ship, which has since been very generally adopted. The type of ship to which Sir F. Beaufort refers was, of course, a sailing man-of-war of the first half of the nineteenth century. As a scientific scale it has one admitted defect. The standard of comparison does not remain the same for all the grades. In all the lower figures up to 4 (inclusive) the *speed of the ship* is the test of the force, in the higher figures it is *the amount of sail which the ship can carry, when close-hauled*, which forms the basis of the classification. This change of standard gives rise to some practical inconvenience, as it is evident that Beaufort's scale does not proceed by equal grades of difference.

However, it seems advisable to give the scale, as it stands a better chance of adoption for use at sea than any other.

The velocities given in the last column have been determined in the Meteorological Office,[1] and are in constant use for the purpose of giving evidence in legal cases. The corresponding velocities, in mètres per second, are obtained by multiplying the values given in the table by the factor 0·447.

Force	Beaufort Scale			Velocity, English miles per hour
0	Calm			3
1	Light air, or just sufficient to give steerage way		.	8
2	Light breeze	or that in which a well-conditioned man-of-war, with all sail set, and, clean full, would go in smooth water from	1–2 knots .	13
3	Gentle ,,		3–4 ,, .	18
4	Moderate ,,		5–6 ,, .	23
5	Fresh ,,		Royals, &c.	28
6	Strong ,,	or that to which she could just carry 'in chase,' 'Full and by' .	Single-reefed top-sails or top-gallant sails	} 34
7	Moderate gale		Double-reefed topsails, jib, &c.	} 40
8	Fresh ,,		Triple-reefed top-sails, &c.	} 48
9	Strong ,,		Close-reefed top-sails and courses	} 56
10	Whole gale, or that with which she could scarcely bear close-reefed main-topsail and reefed foresail			} 65
11	Storm, or that which would reduce her to storm-stay-sails			} 75
12	Hurricane, or that which no canvas could withstand .			90

Inasmuch as scales, varying in greater or less degree from the foregoing, have occasionally been proposed, it may be of interest to quote the following remarks from

[1] *Quarterly Journal of the Meteorological Society*, vol. ii. p. 109.

Schott's discussion of Sir F. Leopold McClintock's observations in the 'Fox,'[1] in which we find a table of pressure and velocity of wind for a scale of ten degrees of force, and Schott says:—

'The relation of the tabular numbers of pressure and velocity is in accordance with Smeaton's Table, and also agrees with that following from D. Bernoulli's formula. By simple proportion, or by means of a diagram, we obtain the following velocity numbers corresponding to Beaufort's scale, or to a graduation from 0—12:—

Force	Velocity	Force	Velocity
0	0	7	40
1	1	8	48
2	4	9	56
3	10	10	67
4	17	11	82
5	24	12	100'
6	32		

It will be admitted that the agreement between this scale and that of the Meteorological Office is close enough for practical purposes over the greater portion of the scale.

There are, however, serious difficulties in the way of instituting a sufficiently accurate comparison between Beaufort's scale and velocities, as instrumentally determined.

Firstly, a clipper ship, which may be considered to correspond with Beaufort's man-of-war, hardly ever comes under the notice of observers on land, or if she does, she will probably have shortened sail when near enough to the shore to be clearly visible; furthermore, the constant use of double topsails detracts in some measure from the value of the test as to reefing.

[1] *Smithsonian Contributions to Knowledge*, No. 146.

THE WIND. 161

Secondly, the estimated force is usually assigned from an observation lasting not more than two minutes, while the corresponding anemometrical velocity is the number of miles of wind which passed the instrument from 30 minutes before, to 30 minutes after, the time of estimation. It must, therefore, necessarily result that, from the gusts and lulls which so frequently occur in the wind, many of the separate observations will have attached to them more, and many less, than the true hourly velocity (whatever that may be) which corresponds to them.

Thirdly, various anemometers have not sufficiently been compared with each other, as already explained.

Fourthly, as stated above, the conditions of exposure of anemometers exercise a very great influence on their action, so that we can hardly compare the data even from adjacent stations to much useful purpose.

We now come to the method of discussion of wind observations—confessedly, one of the most complicated problems in meteorology.

The most usual form of wind records are entries of the direction only, owing to the great difficulty of obtaining accurate reports of the force or of the velocity. If the observations have been taken at regular intervals, the simplest mode of dealing with them is to enter in a table the total number of entries of each individual direction which have been made during each month. Such a table will give us a general idea of the relative prevalence of each direction. If now we want to determine the mean direction from such a table, we may use the formula proposed by Lambert just 100 years ago. It is as follows:—

M

$$\tan \phi = \frac{E - W + (NE + SE - SW - NW) \cos 45°}{N - S + (NE + NW - SE - SW) \cos 45°}$$

In this equation ϕ is the deviation of the mean direction from North, round by East. On the very bold assumption that the force of all winds is equal, the mean motion of the air would be given, in terms of any assumed unit, by the hypothenuse of a right-angled triangle, whose two sides respectively are the numerator and the denominator of the above fraction; but it is evident that any such assumption must often be misleading, for the force of the wind from different points is often very different.

If the force of the wind be recorded, we can find the true mean motion by geometrical construction, it being evident that the result to be sought for is the line AB (fig. 29), which would indicate the ultimate motion of a particle carried over the entire path during the period to which the discussion refers, on the hypothesis that this particle was moved with the same velocity and in the same direction as were respectively recorded by the instrument after it had passed.[1]

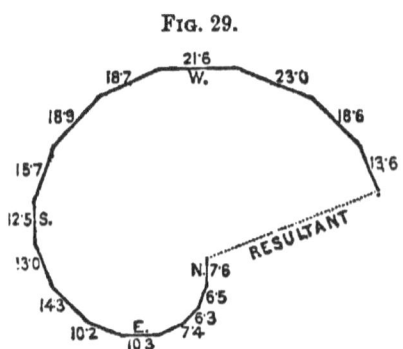

Fig. 29.

Diagram of the mean hourly Velocity of the Wind at Liverpool in January (1852–63).

If we have actual anemometrical records of direction and velocity for each hour during the month, year, &c., we can deduce from them a much closer approximation

---

[1] The diagram is taken from Mr. Rundell's Paper 'On the Winds at Liverpool,' *Quarterly Weather Report*, 1874, App. p. [33].

o the true mean motion of the air than by Lambert's formula, but such a resultant value, given by itself, will afford a very imperfect representation of the real motion which has taken place in the horizontal plane. This is evident from the fact that winds from different points must counteract each other, and so gales of equal force and duration, from opposite quarters, must produce a calm, while the resultant direction may, very possibly, be one from which the wind has hardly ever actually blown during the period under inquiry. Moreover, all the anemometers which have been erected hitherto give indications of the horizontal motion of the air only, and take no account of the upward and downward motions which certainly exist in the atmosphere.

Nevertheless, Lambert's formula is the only one as yet generally adopted, and the mean direction of the wind has been calculated by it for an immense number of places over the northern hemisphere, by the late Professor Coffin, for a paper published by the Smithsonian Institution.[1]

The wind has a very decided diurnal period, as regards its velocity, which is clearly shown by inspec-

Fig. 30.

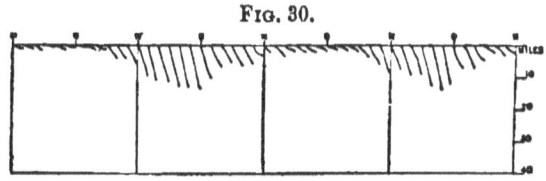

Anemograms at Kew in calm weather, April 16, 17, 1875.

tion of any of the plates of the Quarterly Weather Report during a period of calm weather in summer.

[1] Smithsonian Contribution, No. 268, 'The Winds of the Globe,' by J. H. Coffin, LL.D., with a Discussion, &c., by Dr. A. Woeikof, Washington, 1876.

164   ELEMENTARY METEOROLOGY.

The curve of wind velocity follows a course closely resembling that of temperature, reaching a minimum in fact almost a calm, about sunrise, and a maximun in the afternoon, as is shown in fig. 30 (p. 163). The results of the discussion of the anemograms at Liverpoo for the year 1863[1] give the following figures for *two hourly* velocity in miles, irrespective of direction.:—

		Velocity			Velocity
Midnight to 2 A.M.	.	27·5	Noon to 2 P.M.	.	35·6
2 ,, 4 ,,	.	27·4	2 ,, 4 ,,	.	34·6
4 ,, 6 ,,	.	27·6	4 ,, 6 ,,	.	32·1
6 ,, 8 ,,	.	28·1	6 ,, 8 ,,	.	29·8
8 ,, 10 ,,	.	30·9	8 ,, 10 ,,	.	28·3
10 ,, Noon	.	34·0	10 ,, Midnight.		27·8

This table shows that, at least at Liverpool, there is, on the mean of the year, a decided maximum o velocity between noon and 2 P.M., and that the mini mum falls between 2 and 4 A.M.

As regards the monthly period for velocity, we have results for Armagh, Liverpool, and Sandwick Manse in the Orkneys. They are as follows, in miles per hour :—

	Armagh	Liverpool	Sandwick
January	. . . 13·5 .	. 13·6 .	. 19·1
February	. . . 12·8 .	. 13·8 .	. 20·7
March .	. . . 13·0 .	. 13·2 ·	. 17·4
April .	. . . 11·6 .	. 12·6 .	. 16·3
May .	. . . 7·8 .	. 11·4 .	. 14·2
June .	. . . 4·2 .	. 11·6 .	. 13·7
July .	. . . 6·6 .	. 10·8 .	. 12·0
August .	. . . 7·3 .	. 11·5 .	. 12·1
September	. . . 8·0 .	. 11·2 .	. 13·2
October	. . . 9·1 .	. 12·4 .	. 16·4
November	. . . 10·0 .	. 12·2 .	. 17·3
December	. . . 13·0 .	. 14·0 .	. 20·3
Year	. 9·74 .	. 12·1 .	. 16·1

[1] Mr. Rundell's Paper, quoted p. 162.

As regards the monthly period we see that the ratio between the maximum and minimum is nearly twice as great at Armagh as at Sandwick; at Liverpool there is but little variation comparatively.

It may be worth notice, in passing, to point out that at none of the stations quoted is there any indication of a marked increase of velocity at the time of the equinox, either in spring or autumn, so that the idea of special equinoctial gales being recognisable in these islands is, as far as the above figures indicate, unfounded.

However, all discussions of wind are most laborious, for each individual hourly record must be resolved into its components, and the figures for individual years are so discordant that no period shorter than seven or eight years will give mean values worth having. Usually this resolution has been into two components, North-and-south, and East-and-west respectively, the opposite directions being regarded as positive and negative; but, as Professor Arthur von Oettingen and others have pointed out, such a method is insufficient, and the four components North, South, East, and West must all be given, each being regarded as positive, if we wish to discover the laws of annual and diurnal change of the wind. This is self-evident, because if during a given period, say a week, the wind has blown for six hours from West and for six hours from East, and if the velocities from the two directions have been the same, the resulting motion of the air, as far as these winds are concerned, is *nil*, while there has been a large amount of real movement, which is all masked if only two components are exhibited.

The truth of this statement is illustrated by the

following table, which shows the seasonal values of the components for the year 1873 at Dorpat. The unit referred to is mètres per second:—

	Components				Resultants	
	N.	E.	S.	W.	N.S.	E.W.
Winter	. 0·50	. 0·42	. 1·90	. 2·07	. S. 1·31	. W. 1·6?
Spring	. 1·08	. 0·76	. 1·02	. 1·55	. N. 0·06	. W. 0·7?
Summer	. 0·42	. 0·39	. 0·98	. 1·58	. S. 0·56	. W. 1·1?
Autumn	. 0·32	. 0·59	. 1·63	. 1·54	. S. 1·31	. W. 0·9?

The two columns of so-called 'resultants' give a very incomplete account of the actual movement of the air. In spring, for instance, the resultant motion from North is only 0·06, but when we look to the components we find that this value has been obtained by subtracting a motion from South of 1·02 mètres per second, from a motion from North of 1·08 mètres per second. All traces of this process are concealed if only the final value is given.

The whole problem of wind discussion is of a very complex nature, and its treatment is far from being settled as yet. On the whole, the best mode of publishing the records for different stations is the construction of 'wind-roses.' This is done by calculating the percentage proportion of the number of wind observations from each point of the compass and printing the results either in a tabular form or representing them by a diagram. Wind-roses may be made to show the force as well as the direction, of the wind from the different points.

CHAPTER X.

ELECTRICAL PHENOMENA.

Atmospheric Electricity.—The subjects of hail and of the electricity of the atmosphere are closely connected, inasmuch as hailstorms are always associated with thunder and lightning. The direct proof of the identity of the electricity of a thunder-cloud with that obtainable from an electrical machine was furnished, about 130 years ago, by Franklin's famous experiment with a kite. Three years earlier he had suggested the employment of pointed insulated conductors to test the circumstances on a principle which will be explained at p. 183, but in June 1752, tired of waiting for the erection of such a conductor on a steeple in Philadelphia, he sent up a kite, attaching to the end of the string a key, and to the key a silken ribbon, which he tied to a post under a shed, in order to insulate the apparatus. As soon as the rain wetted the string, and thereby improved its conductivity, sparks were obtained from the key on presenting his hand to it. The experiment was immediately repeated elsewhere with better conducting lines, made of wire, and these yielded sparks or flashes of fire several feet in length. Such experiments are highly dangerous, owing to the very great tension of the electricity engaged, and within three months from Franklin's first experiment a Russian physicist, Professor Richmann, who had arranged an apparatus to bring the electricity into his laboratory, was killed during a thunderstorm. He approached the end of the conducting wire, when a ball of fire apparently leaped to his head, and killed him on the spot.

It is beyond the scope of an elementary book to attempt a detailed description of the apparatus and methods employed in the measurement of atmospheric electricity, and it will suffice to quote the opening paragraphs of the section relating to the subject in the 'Instructions in the Use of Meteorological Instruments.'

'The simplest electroscopes, viz. the Gold Leaf, Bennett's, and Bohnenberger's, are sufficient to show the nature of the electricity present in the air, but it is always found that very little electricity can be observed near the ground, and in order to obtain satisfactory indications the conductor of the electroscope should be brought into contact with the air at some distance from the earth's surface, by means of a collector.

'A simple rough method of collection is to shoot a metallic arrow upwards into the air, the arrow being tied to one end of a conducting string, the lower end of which carries a ring which rests upon the electroscope. The arrow being shot upwards, the electroscope will be found to be electrified as it mounts; and when the ring leaves the plate, the instrument will indicate the state of electrification of the air at that point where the arrow is at the time.

'This manner of observing electricity is simplified by substituting a long conductor reaching upwards. A gilded fishing-rod may be employed, its lower extremity being insulated.

'The usual method employed, however, is Volta's, in which the electricity is collected by means of a flame burning at a height, either in a lantern hung to an insulated mast, and connected to the electroscope by a wire, or by a slow-burning match, attached to the top of a long metal rod. The electricity of the air, in the neighbourhood of the flame, by its inductive action upon the conductor, causes electricity of the opposite nature to accumulate at the upper extremity, where it is constantly carried off by the convection currents in the flame, leaving the conductor charged with electricity of the same kind, and potential, as the air.

'The principle of Volta's method has been made use of by Sir W. Thomson in his water-dropping collector, now employed in observatories, and found to be extremely useful for the observation of atmospheric electricity.

'A copper can is placed on an insulating support, which may be of ebonite, having the surface thinly coated with paraffin; or of glass surrounded with pumice-stone soaked in sulphuric acid. From the can a small pipe projects a considerable distance into the air, and terminates in a fine orifice. The can being filled with water, and the tap which opens into the jet pipe turned on, a small stream of water is allowed to flow out, care being taken that it is so small that it shall break into drops immediately after leaving the nozzle of the tube.

'In half a minute from the starting of the stream, the can will be found to be electrified to the same potential as the air at the point of the tube.

'This collector cannot be employed during the time of frost, unless means are adopted to prevent the freezing of the water in the jet pipe. When observations are to be made with a portable instrument, a slow burning match should be used. Sir W. Thomson recommends for this purpose blotting-paper, steeped in a solution of nitrate of lead, dried, and rolled into matches.

'As to the position of the collector, since electrical density is greater on projecting surfaces, and less on hollow surfaces than on planes, the collector should not be near trees or houses, nor within a closed space.'

Having thus briefly described the instruments, we must proceed to the discussion of the results obtained from their use, and must begin by defining the term *potential*. When we speak of the motion of electricity from one body to another, we say that this is effected owing to difference of potential. Difference of potential in electricity may be said to correspond to difference of level in two portions of fluid in connection with each other, or to difference of temperature in the case of conduction of heat. If one portion of a fluid, as in a river-bed, be at a higher level than another, motion ensues; the water flows down. If one end of a bar be at a higher temperature than the other, heat passes along the bar by conduction. In the same way, when one body is charged with electricity to a higher potential than another, electricity tends to pass off, so as to equalise the potential on the two bodies. In fact, difference of electric potentials may very well be termed 'difference of electric heights.'

Among the principal results which have been obtained with electrometers are the following. When the sky is clear, the potential of the air increases with the height above the ground. This potential is almost always positive.

Quetelet, from the consideration of his five years' observations at Brussels, made the following statements, which have to a certain extent been confirmed by observations made elsewhere, but very few of these observations have really been discussed as yet :—

1. The diurnal march of electricity, at a constant height above the ground, exhibits two maxima and two minima. The maxima fall at about 8 A.M. and 9 P.M. in summer, and about 10 A.M. and 6 P.M. in winter. The day minimum occurs at 3 P.M. in summer and 1 P.M. in winter; the epoch of the night minimum has not been satisfactorily determined.

2. The variations of electricity precede by about an hour those of the barometric range. The maxima occur at the periods of most rapid change of temperature, while the day minimum coincides with the period of maximum temperature and minimum humidity.

3. The annual march of electricity presents one maximum in winter and one minimum in summer, and in Brussels the phenomena were found to be thirteen times more active in January than in June.

It has been stated that the potential of the air is almost always positive, but there are exceptions to this rule, and variations in potential are generally connected with changes in the direction of wind. Clouds may be, and are, electrified, either positively or negatively, and of course the sign of the electricity recorded close to the ground will be affected accordingly. At the time of a thunderstorm the changes in potential and sign of electricity are so violent and rapid that the photographic method of registration employed in Sir W. Thomson's electrograph entirely fails to preserve the record.

As a rule, it may be said that negative electricity indicates rain, while a sudden development of positive electricity in wet weather is a certain sign of the sky clearing.

The subjoined diagram gives an idea of the conditions which accompanied a thunderstorm at Kew Observatory.

FIG. 31.

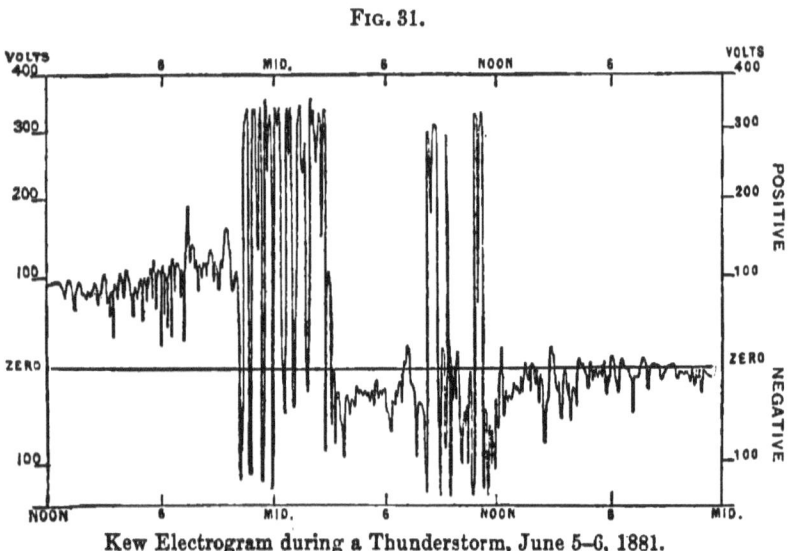

Kew Electrogram during a Thunderstorm, June 5–6, 1881.

Thunderstorms.—Although we are ignorant of the origin of atmospheric electricity, the manifestations of its existence in the phenomena of thunderstorms are familiar to everyone. Thunderstorms are nothing but a series of electrical discharges between cloud and cloud, or between clouds and the earth, and these discharges are manifested in two ways, Lightning and Thunder.

It may seem rather absurd to compare small things with great, but I may say that the greatest distance at which a spark will pass from a powerful electrical machine is about three feet, while lightning flashes are often more than a mile in length, and sometimes extend over four, five, or even ten miles. I shall subsequently (p. 180) return to this subject of the contrast between lightning and the electrical phenomena producible in laboratories.

Of lightning there are three kinds: *zigzag* or *forked lightning*, *sheet* or *summer lightning*, and *globular lightning*, the two first of which are very closely related to each other.

FIG. 32.

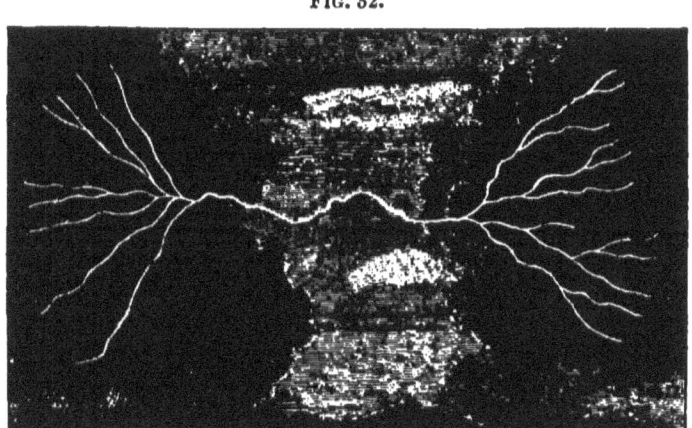

Lightning Flash.

Zigzag Lightning is the discharge which takes place between two oppositely electrified bodies or bodies differing widely in potential, and is precisely similar in character to the spark from an electrical machine, as was pointed out by Franklin long ago.

It follows a zigzag path, owing to the different states of conductivity existing in the air through which it passes, and which offer it an easier path in one direction than in another, probably owing to variations in humidity. At times the flash seems to spread out in a star-shape, or is forked, a single flash apparently giving rise to six or eight, in as many different directions, as represented in fig. 32 (p. 173), copied from Loomis ('Treatise on Meteorology,' p. 171). I have used the term 'apparently,' because it is impossible to say whether a flash of lightning moves from a cloud to the earth, or in the opposite direction.

The flash moves with a velocity which is far too great for the human eye to take in. Sir C. Wheatstone, by his device of a revolving mirror, showed that the duration of a spark, 0·1 in. long, in air, at the ordinary pressure, was about $\frac{1}{24000}$ of a second, and that its velocity along the insulated wire, with which he experimented, was nearly 290,000 miles in a second, being half as great again as the velocity of light— 186,000 miles in a second.

The mind can hardly appreciate such a velocity, but some idea of it is conveyed by the statement that the wheels of a rapidly-moving carriage appear as if at a stand-still when lit up by lightning, and that a disc, painted in segments of various colours, which when rotated rapidly shows, by ordinary illumination, a homogeneous whitish-grey tint, appears, when seen by the light of an electric discharge, to be absolutely at rest, with all its colours distinct.

If the reader will consider what such statements imply, he will see that to say that one flash came down to the earth while another passed up from the earth to a cloud cannot be a true account of the phenomena. The light is sensibly instantaneous along the entire path of the flash. If the spectator is looking at the point on the cloud whence the discharge takes place, the spark *appears to move from the cloud*, because the portion of the flash which is seen directly affects the optic-nerve sooner than the portion seen obliquely.

Sheet Lightning appears as a glow of light illuminating the sky. It is much more common than the preceding form, which is only visible when the discharge takes place in the part of the sky at which the observer is looking. Sheet-lightning and the so-called summer- or heat-lightning are nothing else than the reflection of, or the illumination produced by, distant electrical discharges. Very often the edge of a dark cloud is lit up, showing that the actual lightning flash passed behind that cloud. The most magnificent displays of sheet-lightning are usually coincident with very severe thunderstorms below the horizon, and of which, therefore, the actual zigzag flashes are invisible.

Globular Lightning is a rare phenomenon, and one which has not yet been explained, for it has not as yet been reproduced by electrical machines. It manifests itself as a luminous sphere, varying in diameter from a few inches to even two or three feet. It moves very slowly, and remains visible for several seconds, or even minutes, generally at last exploding with great violence. In the two first particulars it contrasts very strongly with ordinary lightning.

Numerous instances of this phenomenon are on record. Arago, in his Meteorological Essays, cites several reports of it. In one case, at Milan in 1841, one of these globes moved along a street so slowly that the spectators walked after it to watch it, and the narrator saw it from a window, and then ran downstairs, and saw it for three minutes before it struck the cross on a church-steeple and disappeared.[1] Again, a Madame

[1] *Essays*, p. 35.

Espert, in Paris, states that she saw a ball or globe descending from the sky, very like the moon when it appears augmented in size. While she was watching it, 'a terrible explosion burst asunder the envelope, and there darted from it ten or twelve zigzag lightnings, which shot forth in all directions; one of these struck a neighbouring house, where it made a hole in the wall as a cannon-ball might have done .... The phenomenon lasted above a minute.'[1]

To come to experience more recent and nearer home: Dr. Tripe stated to the Meteorological Society[2] that, 'On July 11, 1874, he was watching the progress of the most fearful storm he ever witnessed of hail, rain, wind and lightning, and was looking due south, when he saw a large ball of fire rise apparently about a mile distant from behind some low houses. The ball at first rose slowly, but accelerated its pace as it ascended, so as gradually to acquire a very rapid motion. When it had risen about 45°, it started off at an acute angle towards the west, with such great rapidity as to produce the appearance of a flash of forked lightning. It made three zigzags before it entered the dark cloud.'

The following report of an appearance of globular lightning in the Glendowan Mountains, in the County Donegal, Ireland, by Mr. M. Fitzgerald,[3] is very remarkable:—

'I noticed a globe of fire in the air floating leisurely along. After passing the crown of the ridge, where I first noticed it, it descended gradually into the valley, keeping all the way about the same distance from the

[1] *Essays*, p. 37.
[2] *Quarterly Journal of the Meteorological Society*, vol. ii. p. 431.
[3] *Ibid.* vol. iv. p. 160.

surface of the land, until it reached a stream about 300 yards from where I stood. It then struck the land, and re-appeared in about a minute, drifted along the surface for about 200 yards, and again disappeared in the boggy soil, re-appearing about twenty perches further down the stream; again it moved along the surface, and again sank, this time into the bank of the stream, which it flew across, and finally lodged in the opposite bank, leaving a hole in the peat bank, where it buried itself.

'I at once examined its course, and found a hole about twenty feet square, where it first touched the land, with the pure peat turned out on the lea as if it had been cut out with a huge knife. This was only one minute's work, and, as well as I could judge, it did not occupy fully that time. It next made a trench about twenty perches in length, and four feet deep, afterwards ploughing up the surface about one foot deep, and again tearing away the bank of the stream about five perches in length and five feet deep, and then, hurling the immense mass into the bed of the stream, it flew into the opposite peaty brink. From its first appearance till it buried itself could not have been more than twenty minutes, during which it travelled leisurely, as if floating, with an undulatory motion through the air and land, over one mile. It appeared at first to be a bright red globular ball of fire, about two feet in diameter, but its bulk became rapidly less, particularly after each dip in the soil, so that it appeared not more than three inches' diameter when it finally vanished.'

These instances are sufficient to show that the occurrence of globular lightning is perfectly well established, although some physicists deny the possibility of its existence.

St. Elmo's Fire.—There is a form of electrical display which is manifested during thunderstorms, especially when clouds are low, and which is distinct from lightning. This is the appearance of luminous brushes at the extremities of pointed objects, such as ships, masts, or tree-tops. The phenomenon is of the nature of the 'brush discharge' of the electrical machine, and it receives various names, such as 'St. Elmo's Fire,' and 'Comozants,' a corruption of 'corposants' (*corpus sanctum*). The ancients held its appearance for a sign that Castor and Pollux had come to the seaman's aid, and considered it as an omen of good-fortune. As a fact no damage ever arises directly from this display of electricity. At sea it is not uncommon, during thunderstorms, on the mast-heads and yard-arms. Clavering and Sabine, in H.M.S. 'Griper,' in a gale off Trondhjem, in Norway, in the winter of 1823, saw as many as eight of these flames in various parts of the rigging.

Thunder is the noise which accompanies the electrical discharge. The sound is produced by the sudden dilatation of the air along the track of the spark. Air, being a non-conductor, resists the passage of the electricity, and so heat is generated, which dilates it enormously along the path, and, the instant this heating action ceases, the expanded air contracts again with great violence. This gives rise to a sudden clap or explosion, and when lightning passes close to the observer, as when it is said to strike an object in his neighbourhood, a sudden crash is heard, with little rolling noise. The pealing or rolling of thunder is due in great measure to echoes, and of this anyone in a mountainous country can satisfy himself by hearing the repeated echoes of a single shot. The surfaces from which the echoes are reverberated are the surfaces either of clouds or of masses of air of very different densities, which are known to exist in proximity to each other when thunderstorms are prevalent. These have been proved by Professor Tyndall to interfere with the direct transmission of sound, and to reflect it, as if they formed solid walls. Another cause is well stated by Professor Tait in his recent lecture on 'Thunderstorms:'[1]—'When the flash is a long one, all its parts being nearly equidistant from the observer, he hears the sound from all these parts simultaneously; but if its parts be at very different distances from him, he hears successively the sounds from portions farther and farther distant from him. If the flash be much zigzagged, long portions of its course may run at one and the same distance from him, and the sounds from these arrive simultaneously at his ear.' Thunder is inaudible even at moderate distances, for there appears to be no evidence of its being heard further off than about fifteen miles, while the report of a cannon has been heard fifty miles off, and the noise of a bombardment or a naval engagement at double that distance.

As to the estimate of the distance of the flash from the observer, by means of the interval of time which elapses between it and the thunder which follows, an allowance of about five seconds to the mile is near enough.

The reason that the electrical phenomena mani-

[1] *Nature*, vol. xxiii., p. 409.

fested in the atmosphere are so incomparably more intense than any we can reproduce in our laboratories is in a general way explained on the supposition that the electricity is first developed on the original molecules of water as they are formed by condensation of vapour. These molecules combine to form cloud particles or droplets, and these latter ultimately to form raindrops. Professor Tait goes on to say on this subject:—'If the vapour particles were originally electrified to any finite potential, the cloud particles would be each at a potential enormously higher, and the raindrops considerably higher still. . . . . Now, the number of particles of vapour which go to the formation of a single average raindrop is expressed in billions of billions; so that the potential of the drop would be many thousands of billion times as great as that of a particle of vapour. On the very lowest estimate this would be incomparably greater than any potential we can hope to produce by means of electrical machines.'

However, this subject has not as yet received a complete explanation, and we must only wait for further researches upon the subject.

I am, however, indebted to the kindness of Mr. De La Rue for the following calculation of the potential necessary to produce a flash of lightning a mile in length. By his and Dr. Müller's experiments ('Philosophical Transactions,' vol. clxix., page 118) with his magnificent battery, the striking distance, between points, when 11,000 cells were used, the potential of each being 1·06 'volts,' was 0·62 inch. This striking distance varies with the square of the number of cells employed. Then, as 1 mile = 63,360 inches, we

have $\sqrt{\dfrac{63,360}{0\cdot 62}} \times 11,000 = 3,516,480$ cells
as the amount requisite to produce such a flash.

It may be useful to say something about the damage done by lightning and the modes of protection against it. Not only are animals killed, and buildings, &c., set on fire, but trees are rent in pieces, and heavy blocks of masonry, like chimney-stacks and turrets, thrown down. These effects strongly resemble those produced by electrical batteries and machines, but are much more intense in their character and in their physical and physiological action. In the case of animals killed the electricity produces a violent excitement of the muscular and nervous tissues, occasioning death, frequently without external wounds.

In the case of trees the disruption is attributed to the sudden volatilization of the sap and expansion of air in the vessels, by the heat generated by the passage of electricity through these badly conducting substances. The disintegration of stones has been attributed by Pouillet to the effect of induction, which takes place instantaneously, and with such energy that the particles of these imperfect conductors are unable to accommodate themselves to the new distribution of electricity with sufficient rapidity.

There is, however, one action of lightning which calls for special notice. This is termed the 'return shock,' and persons may be killed by it without the actual passage of a flash between their bodies and an electrified cloud. The explanation is as follows :—

If a cloud positively electrified to a high potential pass over a place, the air beneath it, being a di-electric, becomes polarized and causes a large quantity of electricity of the opposite kind to accumulate on the earth's surface immediately below it. This inductive charge is shared by all objects on the ground. When therefore the cloud becomes discharged either by a flash to another cloud or to the earth, the induced electricity at once disappears. In so doing it will produce a nervous shock in any living object subject to its influence, which will probably prove fatal if the original charge of electricity is strong.

I am indebted to Professor W. Grylls Adams, F.R.S., for the following account of a curious instance of the inductive action of thunderstorms which occurred to himself and others on the Great Aletsch Glacier, in the Bernese Oberland, on July 10, 1863. 'At the top of the Jungfrau Joch at 10.5 A.M. the party were met by a violent hailstorm which came rolling over from the northern side of the col. After walking for two hours down the centre of the Aletsch Glacier with the thunderclouds gathering around, sounds similar to the singing of a kettle began to issue from the tops of the alpenstocks and pricking sensations were felt at the top of the head, and other well-known evidences were experienced that a process of electrical charging of insulated conductors was going on. The process of charging or discharging went on for some minutes with gradually increasing intensity by a kind of brush discharge and was then suddenly terminated by an electric shock, when the clouds above discharged themselves to one another. Accompanying these sensations there were pricking sensations at the waist where the rope passed from one of the party to the next, showing that as condensers of electricity they were not all of the same capacity or all equally affected by the same charge. As a rule, the

ELECTRICAL PHENOMENA. 183

youngest showed signs of being most affected, although one gentleman who had his arm paralysed for a time was not the youngest of the party.

'The process of slow and gradual charging, followed by a sudden discharge, giving a severe shock whenever there was a peal of thunder overhead, was repeated over and over again several times until after about twenty-five minutes the storm passed away. These facts point to a direct relation between the temperament of the individual and his capacity of being excited electrically or his specific inductive capacity.'

A mode of protection against lightning was devised by Dr. Franklin, who proposed the Lightning-conductor now universally adopted, and which consists of a pointed rod of good conducting material, such as copper, placed in the immediate neighbourhood of the object to be protected and in efficient connection with the ground. The action of such a conductor allows the electricity of the earth to flow off quietly, and gradually neutralise the electricity in the atmosphere about, so as to prevent the accumulation of such a charge as would give rise to a lightning-flash. The action depends on what is called the 'power of points.' The electricity on a sphere is uniformly distributed over the surface; on an oval figure it tends to accumulate at the ends. On a cylinder this tendency is more strongly developed, and, when the cylinder becomes a fine wire, the tension is so great at the end that the electricity soon forces its way into the surrounding air and escapes.

The following account, abridged from one printed in the 'Quarterly Journal of the Meteorological Society,' vol. ii., p. 423, will give some explanation of this action of conductors:—In August 1865 a Mr. F. G. Smith, with three companions, ascended the Piz Languard, in the Engadine, in Switzerland, and was caught in bad weather. They, however, reached the summit, 11,000 feet high, which is a sharp narrow ridge, having a flag-staff at one end and at the other a dial-plate, marked with the bearings of the surrounding peaks, &c., mounted on stone and covered with an iron hood. When they stood upon the ridge they were enveloped by the cloud and falling snow; but soon they heard a strange crackling sound, like the rattling of hail-stones against a window. Investigation into the cause of this noise showed that it proceeded from the flagstaff— sometimes from the top, sometimes from the bottom— varying in intensity, but never ceasing. The party held up their alpenstocks, points downwards, and then distinctly felt the electrical currents passing through their bodies and heard the crackling as they passed into the sticks. On turning the points upwards the crackling sound grew louder and the thrilling sensations became much more powerful. They then experienced the sensation very strongly in their temples and at the finger ends. The discharge lasted the whole period the party stayed on the summit—about three-quarters of an hour—and during that time there was no electrical discharge whatever from the dial-plate or hood, which were not sharp-pointed objects.

If we have got a pointed conductor projecting well above a building which is to be protected, we must take care that the section of the conductor is sufficient to allow the electricity to pass along it, and that the rod is perfectly continuous. If the section be insufficient, the electricity may either heat the conductor until it melts and breaks—the experiment of melting a thin wire, by passing a strong electrical current along it, is a familiar one—or it may leap from the conductor to some other channel which affords an easier passage to the ground. If the conductor be not continuous it is quite useless, for the electricity will strike across the gaps in the connection, and may do great damage. It is, therefore, necessary to examine all joints in the conductor from time to time, for if these are corroded the continuity is wholly, or in part, interrupted.

Finally, it is of paramount importance to see to the connection with the ground. A metallic rod is a much better conductor than the earth, and, accordingly, if we simply plunge the end of such a rod into the ground, the result will be just the same as if we were suddenly to reduce its section. As the amount of water flowing through a pipe is regulated by the section of that pipe at its narrowest part, so the amount of electricity flowing along a conductor depends on the amount which can pass through the part of the conductor which offers the most resistance to its passage. The most efficient mode of making earth-contacts is to connect the rod with a mass of metal affording a large surface in contact with moist earth. In a town this is best done by connecting it with the gas and water mains in the streets. Other modes are to make the conductor end in a sort of harrow, or cage, packed in coke, which soon gets soaked with moisture; a length of chain buried in coke serves the same purpose.

The best form of conductor is a rod of copper: copper-wire ropes are also good, but are more liable to corrosion. On the Continent, copper bands are generally used, and there is not very much difference between these and rods. A rod ½ inch in diameter is amply sufficient for all ordinary buildings. Iron can also be used for the material of the rod, but it is liable to rust, and it must present to the electricity five times the sectional area that copper need offer, as its conducting power is only one-fifth as great as that of copper.

The indispensable conditions for the efficient protection of buildings from lightning are given in a paper by Dr. R. J. Mann, in the 'Quarterly Journal of the Meteorological Society,' vol. ii., p. 428. The Society, however, at a subsequent date organised a conference of delegates from various scientific and professional societies to examine into the question.

The Code of Rules laid down by this Conference will be found in the note.[1] These do not materially differ from those proposed in 1875 by Dr. Mann, but the Report itself presents us with a mass of valuable evidence collected by the members, and a most carefully compiled bibliography of the entire subject.

[1] CODE OF RULES FOR THE ERECTION OF LIGHTNING-CONDUCTORS. *Report of the Lightning-rod Conference.* London: Spon. 1882. Pp. 16.

Points.—The point of the upper terminal should not be sharp—not sharper than a cone of which the height is equal to the radius of its base. But a foot lower down a copper ring should be screwed and soldered on to the upper terminal, in which ring should be fixed three or four sharp copper points, each about 6 inches long. It is desirable that these points should be so platinized, gilded, or nickel-plated as to resist oxidation.

Upper Terminals.—The number of conductors or points to be specified will depend upon the size of the building, the material of which it is constructed, and the comparative height of the several parts. No general rule can be given for this; but the architect must, however, bear in mind that even ordinary chimney-stacks, when exposed, should be protected by short terminals connected to the nearest rod, inasmuch as accidents often occur owing to the good conducting power of the heated air and soot in a chimney.

Insulators.—The rod is not to be kept from the building by glass or other insulators, but attached to it by metal fastenings.

Some explanation may be advisable as to the necessity of keeping the conductor clear of internal gas-

Fixing.—Rods should preferentially be taken down the side of the building which is most exposed to rain. They should be held firmly, but the holdfasts should not be driven in so tightly as to pinch the rod, or prevent the contraction and expansion produced by changes of temperature.

Factory Chimneys.—These should have a copper band round the top, and stout, sharp, copper points, each about 1 foot long, at intervals of 2 or 3 feet throughout the circumference, and the rod should be connected with all bands and metallic masses in or near the chimney. Oxidation of the points must be carefully guarded against.

Ornamental Ironwork.—All vanes, finials, ridge ironwork, &c., should be connected with the conductor, and it is not absolutely necessary to use any other point than that afforded by such ornamental ironwork, provided the connection be perfect and the mass of ironwork considerable. As, however, there is risk of derangement through repairs, it is safer to leave an independent upper terminal.

Material for Rod.—Copper, weighing not less than 6 oz. per foot run, and the conductivity of which is not less than 90 per cent. of that of pure copper, either in the form of tape or rope of stout wires, no individual wire being less than No. 12 B. W. G. Iron may be used, but should not weigh less than $2\frac{1}{4}$ lbs. per foot run.

Joints.—Although electricity of high tension will jump across bad joints, they diminish the efficacy of the conductor; therefore, every joint, besides being well cleaned, screwed, scarfed, or riveted, should be thoroughly soldered.

Protection.—Copper rods to the height of 10 feet above the ground should be protected from injury and theft by being inclosed in an iron pipe reaching some distance into the ground.

Painting.—Iron rods, whether galvanised or not, should be painted; copper ones may be painted or not, according to architectural requirements.

Curvature.—The rod should not be bent abruptly round sharp corners. In no case should the length of the rod between two points be more than half as long again as the straight line joining them. Where a string course or other projecting stonework will admit of it, the rod may be carried straight through instead of round the projection. In such a case the hole should be large enough to allow the conductor to pass freely, and allow for expansion, &c.

Extensive Masses of Metal.—As far as practicable it is desirable that the conductor be connected to extensive masses of metal, such as hot-water pipes, &c., both internal and external; but it should be kept away from all soft metal pipes, and from internal gas-pipes of every kind. Church bells inside well-protected spires need not be connected pipes. The danger is that at any time either the points or the earth-contact may become deranged or impaired in efficacy, so that a discharge of high potential may pass, instead of a continuous stream of low potential, may make its way into such channels, and in so doing form a spark which may ignite the gas.

The appearance of the sky during a thunderstorm is sufficiently familiar to most persons. The most characteristic feature is a heavy cumulus cloud, with an extensive development of cirro-stratus above and a sort of cloud curtain of loose texture underneath, which partly hides the falling rain or hail. The cumulus cloud is often very solid and opaque, at times producing darkness equal to that of the darkest London fogs. That this is merely caused by the obstruction of daylight is clearly proved by the distance at which objects lit up by lightning can be seen during a thunder-storm. On a recent occasion, in Switzerland, the writer saw snow-peaks, distant over fifty miles from

Earth Connection.—It is essential that the lower extremity of the conductor be buried in permanently damp soil; hence proximity to rain-water pipes, and to drains, is desirable. It is a very good plan to make the conductor bifurcate close below the surface of the ground, and adopt two of the following methods for securing the escape of the lightning into the earth:—A strip of copper tape may be led from the bottom of the rod to the nearest gas or water *main*—not merely to a lead pipe—and be soldered to it; or a tape may be soldered to a sheet of copper 3 feet × 3 feet and $\frac{1}{16}$ inch thick, buried in permanently wet earth, and surrounded by cinders or coke; or many yards of the tape may be laid in a trench filled with coke, taking care that the surfaces of copper are, as in the previous cases, not less than 18 square feet. Where iron is used for the rod a galvanized iron plate of similar dimensions should be employed.

Inspection.—Before giving his final certificate the architect should have the conductor satisfactorily examined and tested by a qualified person, as injury to it often occurs, up to the latest period of the works, from accidental causes, and often from the carelessness of workmen.

where he was standing, quite clearly, when flashes appeared in the direction in which they lay.

Thunderstorms are, speaking generally, associated with great differences of temperature in adjacent masses of air. These conditions are especially likely to occur in hot climates, where the soil gets excessively heated during the day-time, and also in cooler climates in winter, where sudden changes of temperature characterise the storms of the season. Professor Mohn classifies thunderstorms under two heads—*heat thunderstorms* and *cyclonic thunderstorms*. The former are the type which are predominant in summer and in hot climates. The latter are characteristic of our own Atlantic coasts, as well as of Iceland and Northern Norway, they accompany cyclonic disturbances, and, in fact, are reported as a common feature of our winter gales of wind. They are more dangerous than summer storms, though they are not so violent, for, as the clouds drift at a lower level, the lightning more frequently strikes the ground than in heat thunderstorms.

It is easy to see why this class of thunderstorm is more frequent on our Atlantic seaboard than the other, because, under the conditions of climate to which the proximity of the sea gives rise, the lower strata of the atmosphere can never become so intensely heated as is the case inland, and consequently great contrasts of temperature, along a vertical section of the atmosphere, cannot easily arise.

In illustration of the statement that thunderstorms are associated with sudden and serious alterations of temperature we may observe, firstly, that cumulus is the predominant type of thunder cloud, which type indicates the presence of a colder and drier stratum of air lying immediately above the saturated stratum in which the cloud is formed. Secondly, that hail hardly ever falls except during thunderstorms. Hail requires for its production the presence of a mass of air at a very low temperature, while, as it usually occurs during the warmer portion of the year, the lower layers of the atmosphere are relatively hot when it falls. Thirdly, we may notice that thunderstorms are always associated with a change of wind. They generally come up in a direction differing from that of the surface wind, and in the course of the storm the wind shifts to a certain extent—usually the sudden shift of wind in a cyclonic storm towards north-west (p. 359). That sufficient changes of temperature may occur in winter storms to generate thunderstorms will readily appear when it is remembered that an almost instantaneous fall of the thermometer of ten, or fifteen, degrees, or even more, is not an uncommon accompaniment of this shift of wind even in our temperate climates. In extreme climates, such as those of Russia, or, more particularly, of the Western States of North America, even greater changes occur.

The way in which changes of temperature tend to produce electrical disturbance is not as yet explained, but it is undeniable that they form an essential condition for the generation of a thunderstorm.

Thunderstorms exhibit a diurnal and an annual period in the frequency of their occurrence. The summer storms are most frequent at the hottest period of the day—the early afternoon. The winter storms occur at all hours, as the cyclonic storms with which they are connected do not exhibit a very marked diurnal period in the time of their passage over each station.

ELECTRICAL PHENOMENA. 191

The existence of a yearly period has already been indicated when speaking of the two types of storms. It is strongly marked in high latitudes, for all the thunderstorms in Iceland occur in winter.

As to the geographical distribution of thunderstorms, they are far more abundant in hot than in temperate climates. In fact, it was formerly stated, on the authority of Arago, that they were unknown in the Arctic Regions, and they really occur there with extreme rarity. However, as the Norwegian Captain, Johannsen, had one, in 78° N. lat., at Bell Sound, Spitzbergen, August 23, 1873, and as a succession of thunderstorms was reported for several days in July 1870, on the west coast of Nova Zembla, the occasional occurrence of the phenomenon in the Arctic Regions is indisputable.

On the other hand, we have already (p. 146) cited Caldcleugh's statement that at Rio, in issuing invitations for afternoon parties, it was usual to specify whether the guests were to assemble before or after the thunderstorm, which was a daily phenomenon. In Abyssinia, d'Abbadie gives, on the average of four years, 410·6 as the annual number of thunderstorms; many of these, however, consist of only one or two flashes of lightning. The daily period in this region is so marked that, out of 1,909 storms recorded in six years, only 22 occurred between midnight and 11 A.M.

The Aurora.—The aurora is a phenomenon which is evidently electrical in its character, and one the nature and origin of which are as little understood as are those of atmospheric electricity. As seen in these islands, it most usually consists of a luminous display on the north-western horizon—a low arch of light is visible, under which the sky appears to be darker than the rest of the heavens where they are unclouded. This is the so-called 'dark segment' of the aurora. Multiple arches are sometimes seen. On one occasion, at Bossekop in Finmark, Lottin counted as many as nine, simultaneously visible.

The luminous display frequently develops itself into shafts or streamers which shoot upwards irregularly, giving an appearance of waves of light passing over the sky. The direction of these streamers in these latitudes is always strictly parallel to the dipping needle, so that they are directed, not to the zenith, but to the point towards which the south pole of such a needle points. Occasionally, the streamers collect round this spot and form a luminous ring, the so-called 'corona' of the aurora, which, however, is only a transient phenomenon, for as the flow of streamers which have produced it fades away the appearance dies out.

The whole display resembles a curtain of light hanging parallel to the dipping needle, and when the elevation of such a curtain is so great that it is over our heads and we see only the lower edge, we have the phenomenon of the corona.

The colour of the display is usually white, but the arch and streamers occasionally assume other tints, red being the most usual variation; a red aurora is especially striking, as the sky appears lit up by a vast conflagration.

There is one feature of auroral phenomena about which there is a great conflict of evidence. Several trustworthy observers assert that a peculiar crackling noise is heard during the display, while most scientific men who have taken part in Arctic explorations deny that any sound whatever is perceptible. It seems not impossible that this conflict of testimony between experienced observers may be due to the different sensibilities of different persons to very high notes.

As to the height above the ground at which the display takes place, there is great uncertainty. At times the streamers are close to the earth's surface. Thus in 1836, Sir E. Sabine, when at anchor in a yacht in Loch Scavaig, in Skye, saw a thin mist enveloping a hill about 2,000 feet high, to the east of the loch. This cloud became luminous at dusk, though still so thin as to allow the outline of the hill to remain visible through it, and at night streams of the aurora ascended from it.

Occasionally, one observer has noticed an aurora near the zenith, while another at no great distance has observed the light close to the horizon. From observations like these the elevation has been estimated at a few thousand feet.

Messrs. De La Rue and Müller have submitted to the Royal Society some results which they have obtained and some calculations they have made as to the height at which the display may probably occur, and as to the tint, &c., which is visible ('Proceedings of the Royal Society,' vol. xxx. p. 332). In each case Mr. De La Rue's battery of 11,000 cells was employed.

The colour of the discharge, with the same potential, is found to vary greatly with the tenuity of the gas or air. Thus at a pressure corresponding to a vertical height of 12·4 miles, the discharge has the carmine tint so frequently observed in the display. At a pressure corresponding to a height of 30·86 miles, the discharge becomes salmon-coloured. At a pressure corresponding to 33·96 miles, the tint is of a paler salmon-colour, and as pressure is still further reduced it becomes of a pale milky-white. The roseate and the salmon-coloured tints are always in the vicinity of the positive source of the electric current. The discharge at the negative terminal, in air, is always of a violet hue, and this tint in the aurora therefore indicates a proximity of the negative source.

The authors give a table of the pressures at which they actually obtained discharges, with the corresponding calculated heights, and also of the character of the light in each case, from which the following figures are taken :—

Height in miles	Tint	Height in miles	Tint
81·47	Pale and faint	27·42	Carmine
37·67	Maximum brilliancy	17·86	,,
33·96	Pale salmon	12·42	,,
32·87	Salmon coloured	11·58	Full red
30·86	,, ,,		

They conclude the paper with the words—'It is conceivable that the aurora may occur at times at an altitude of a few thousand feet.'

The distribution of the aurora on the earth's surface has been studied carefully by Loomis.

In the northern hemisphere it appears most frequently, not at the north pole, but in a belt which is oval in shape and surrounds the pole. On the side of Russia and Siberia it extends from the Arctic circle to 75°. In America it lies much farther south and is broader, stretching from 50° to 62°. North of this band auroras are less frequent and when visible, are usually seen to the southward of the observer. Over the belt of maximum frequency more than eighty auroras are seen annually. The course of the belts of equal auroral frequency bears some general resemblance to that of the isoclinal lines or curves of equal magnetic dip.

The aurora is by no means confined to the northern hemisphere. It is rarely seen close to the equator, or in low latitudes generally, but it appears just as frequently in the southern sky as in the northern, and of course there receives the name of Aurora Australis. Several especially brilliant displays, of late years, have been noted simultaneously in Europe and in Australia, and in fact there is every reason to believe that all northern auroras must have their corresponding southern displays.

As to the connection between the appearance of the aurora and weather nothing very positive can be asserted. That there is some relation appears from the circumstance that at the time of a brilliant display the sky is constantly being covered with cloud and clearing again. In these islands, at least, an aurora is rarely seen except at times of atmospheric disturbance; in fact it not unfrequently accompanies severe storms. When, however, we pass to the region of greater auroral frequency no such agreement is observable, and most Arctic travellers deny its existence. In fact, the late Lieut. Weyprecht, the Austrian explorer, stated that after discussing carefully the observations taken during his long sojourn in high latitudes, on the voyage in which Franz Josef Land was discovered, he could find no trace of a connection between the aurora and the weather.

As to the connection between the aurora and the phenomena of terrestrial magnetism much remains to be explained. It is generally stated that during the continuance of an aurora the indications of the magnetic instruments are much disturbed, that, in fact, what is called a 'magnetic storm' occurs. This would be accounted for if the aurora were a discharge of an electrical current in the upper regions of the atmosphere, for, as is well known, the flow of an electric current near a magnet at once affects its magnetic condition. However, it appears from the continuous magnetic records at Kew that all auroras are not coincident with disturbances of the magnetic instruments; and Dr. Wijkander, in his discussion of the magnetic results of the Swedish expedition to Spitzbergen, in 1871, states that, as to intensity, no relation is traceable between the magnetic disturbances and the brilliancy of auroral phenomena. The subject, therefore, demands fuller inquiry.

Ozone.—Before leaving the subject of electrical phenomena we must not omit all mention of ozone, although in the opinion of most meteorologists at present, the less said about that substance the better.

Ozone was discovered by Schönbein in 1848, and he gave it its name from its peculiar odour (from the Greek word ὄζω, I smell). We often recognise a peculiar odour in the atmosphere after a thunderstorm, and this is popularly taken, in the case of lightning striking near a person, for the smell of burning sulphur! One of the modes of production of ozone is the passage of electric sparks through a confined portion of air, and this accounts for its presence in thunderstorms.

Ozone is nothing else than oxygen in what chemists call an *allotropic* condition; it still remains an invisible gas, but is endowed with properties different from those of ordinary oxygen. It has become much more energetic in its chemical action, oxidising objects exposed to it with great rapidity. It is a good disinfectant of the atmosphere, inasmuch as it hastens the oxidation of decomposing animal and vegetable matter: hence it is important to life and health.

It is more abundant on the sea-coast than inland, in the country than in towns, and it is stated to accompany South-westerly winds as distinguished from North-easterly.

The mode of observing it is to expose slips of white blotting paper, soaked in a solution of iodide of potassium and starch, for a definite period, in a cage allowing free access of air, while excluding direct sunlight and rain, and to note according to a definite scale the degree of discoloration the paper exhibits.

Having said thus much about the substance, I give the reasons for the opinion expressed above—reasons so well founded that at the recent International Congresses at Vienna in 1873, and at Rome in 1879, it was unanimously agreed that no mode as yet suggested of observing ozone could be recommended for adoption.

In the first place, a substance, called the peroxide of hydrogen, is occasionally present in the atmosphere. This exerts an action similar in many respects to ozone, for which it is often mistaken, but it does not possess the same beneficial qualities. Not only this gas, but other oxidising agents, such as nitric acid, may affect the ozone papers. In fact, in the course of a recent discussion on ozone, before the Meteorological Society,

Dr. J. W. Tripe, an experienced observer of ozone, stated that the highest degree of discoloration which he had ever noticed had occurred after a grand display of fireworks in his neighbourhood. This, therefore, was not caused by normal atmospheric action.

Again, it is evident that if we are to compare the proportion of ozone in the atmosphere on different days we must cause precisely the same amount of air to pass through the cage on each day. But simple exposure to the air will not effect this, as the wind varies in force continually.

Lastly, the discoloration of the paper is not permanent; the tint once developed may be again discharged by atmospheric action, so that we can have no certainty that we find recorded on the paper the real amount of ozone present in the air.

CHAPTER XI.

OPTICAL PHENOMENA.

THERE are four classes of phenomena depending on the action of light: the *rainbow*, the *corona*, the *halo*, and the *colours of the sky and clouds*.

A *rainbow* is produced whenever the observer is in full sunshine, while rain is falling on the side opposite to that in which the sun is situated. The bow is a portion of a circle whose centre lies on the prolongation of a line from the sun to the observer's head, the radius of the red bow being nearly 42°. The amount of the circle visible at any time depends on the altitude of the sun. If the sun be actually on the horizon, the bow will be a semicircle. If the sun be at an altitude of 42°, the bow will be depressed, and the top of the red arch will just touch the horizon. With greater altitudes of the sun the bow will be beneath the horizon altogether, and therefore invisible. For this reason the rainbow is more commonly seen in the morning and evening than in the middle of the day, especially in summer.

When the observer is situated on a mountain top, the bow sometimes forms an entire circle, and we all know that complete circles are occasionally seen in the spray of a waterfall.

It is hardly necessary to describe the bow. It exhibits the prismatic colours in their well-known order, the red being outside, the violet inside. This is the *primary bow*. Frequently we see outside it another, the *secondary bow*, presenting the prismatic colours in the reverse order to that just described. The radius of the outside arch, the violet in this case, is almost exactly 54°.

The explanation of all these phenomena belongs to optics. They arise from the reflection and refraction of the rays of light from the inner surfaces of, and through, raindrops.

The primary bow is caused by one reflection and two refractions.

The secondary bow is caused by two reflections and two refractions.

We frequently notice inside the violet arch of the primary, and outside the arch of the secondary, bow, a series of coloured bows, alternately red and green. These are called 'supernumerary' bows, and arise from what is called, in optics, the 'interference' of rays of light.

Occasionally the condition of the atmosphere is such that the more refrangible rays of light are stopped, and all that remains of the seven colours is the red, giving the appearance of a red bow.

The appearance of fragmentary bows, 'wind-galls,' 'weather-galls,' or 'wind-dogs,' as they are popularly called, is generally supposed to be a sign of storm. More correctly they are the effect of a storm, for they arise in showery weather, when the clouds are very much broken up, so that a sufficiently wide sheet of falling rain is not present to form the complete bow. Fragmentary bows are frequently to be seen in the spray of a waterfall, of a breaking wave, or even of a fountain.

Occasionally what are termed 'extraordinary' bows are seen, when the sun's rays are reflected, say from a water surface. If these conditions produce a distinct image of the sun, this, in its turn, shining on the raindrops can produce a bow which will not coincide with that formed by the direct rays. At times these bows are placed even laterally with regard to the ordinary bow. Such a phenomenon was witnessed at Kirkwall, Nov. 13, 1871.[1]

The phenomenon of a white rainbow is sometimes seen on a thick fog. Such bows are called 'fog-bows,' and in some parts 'fog-eaters,' as they are a sign that the fog will soon disappear, under the action of the sun's heat. The explanation of their production is that the drops of water which float in the air, and form the fog, are extremely small. The smaller the drops the broader is the bow they produce, and this finally appears as a faintish-white arch, without any separation of the prismatic colours. The diameter of the fog-bow varies between 38° and 41°.

As the rainbow always appears on the opposite side of the sky to that in which the sun is situated, and in the direction in which rain is falling, it is evident that a rainbow seen in the forenoon must be in the west, and one in the afternoon in the east. The popular saying, therefore,

> A rainbow in the morning
> Is the shepherd's warning;
> A rainbow at night
> Is the shepherd's delight,

simply implies that if it is raining to the west of the observer in the morning, that rain will probably pass

[1] *Quarterly Journal of the Meteorological Society*, vol. i. p. 237.

over him during the day, while if it is raining to the east of the observer in the evening the sky to the westward must be clear, and the night will probably be dry. The justification for the saying depends on the fact that in these latitudes weather changes generally move from west to east.

Lunar rainbows are produced in precisely the same way as solar, but as the light of the moon is much fainter than that of the sun, the bow is not nearly so brilliant, and so may frequently escape notice. It often appears merely as a whitish or yellowish arch. It is also of much rarer occurrence than the solar bow, for it is only when the moon is nearly full that she gives light enough to produce the phenomenon.

Coronæ are the coloured rings we see round the sun or moon when thin clouds pass across them. They are frequent enough in the case of the moon, but the sun's light is too blinding for us to recognise their appearance in the daytime unless we look at him through smoked glass, or by reflection in a black mirror or in still water, in order to reduce the brilliancy.

The rings show prismatic colours, and the tints are arranged in the same order as that which obtains in the rainbow; the violet being inside and the red outside.

The phenomenon can be reproduced by looking at the moon through a plate of glass dusted over with lycopodium seed. It is due to what is called '*diffraction*,' to the deflection of the rays of light in passing beside the minute droplets of water which form the cloud. The smaller the size of these globules the greater is the diameter of the corona, and so when the rings are contracting it is a sign that the globules are growing in size, that condensation is making further progress, and that rain will shortly fall.

Glories, or 'anthelia,' are coloured circles seen round the shadow of the observer's head when thrown on a fog. Scoresby observed the phenomenon frequently in the Arctic Regions, and says that it was always visible when the sun shone on a fog. Glories have been seen round the shadow of a balloon thrown on a stratum of clouds beneath it.

Halos are the large circles seen round the moon, and less frequently round the sun, when cirro-stratus clouds exist. Generally the phenomenon is colourless, presenting itself as a white ring, but occasionally prismatic colours are developed, the red being inside ; the arrangement opposite to that of the corona.

The halos which are commonly observed are of two kinds ; one, the more usual, of 22° radius, the other of 46°. Occasionally supernumerary circles appear, and near their points of intersection with the primary halo, the illumination is more brilliant, and what are called 'mock suns' (*parhelia*), and 'mock moons' (*paraselenæ*), are developed. The commonest position for these phenomena is on the same level as the luminary, and distant 22° from it on the east or west.[1]

Halos are due to refraction. They would be theoretically producible if the rays were refracted through minute crystals of ice floating in the air in all

---

[1] By theory, the parhelia (for instance) differ in azimuth from the sun by an angle somewhat greater than the radius of the halo, except when the sun is on the horizon. Accordingly, except under the circumstances just stated, the parhelia lie a little outside the halo, and so much the more the higher the sun is. The same reasoning of course applies to the moon.

204 ELEMENTARY METEOROLOGY.

sorts of positions. As the level at which the cirrostratus, the halo-forming cloud, floats is over 20,000 feet, and as at this level the temperature must be far below the freezing-point, we are justified in assuming

Fig. 33.

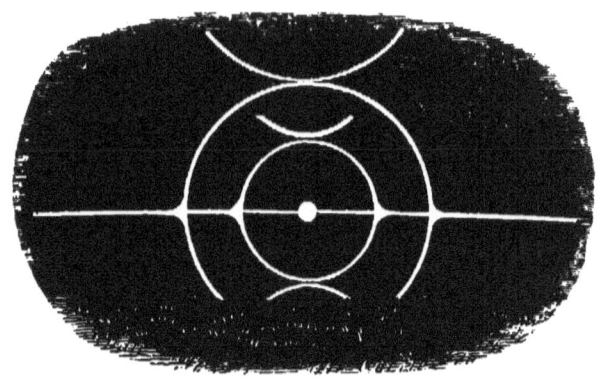

Halo with Parhelia (Loomis).

that these clouds are composed of minute ice spiculæ, not of droplets of water, and that the theoretical explanation of the production of the halo is correct. We are, however, not left entirely to theory, for several observers (Sir E. Parry at Port Bowen, Kämtz on the Faulhorn, &c.) have seen halos, and at the same time spiculæ of ice floating in the air.

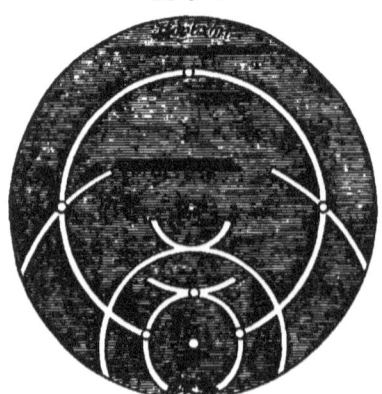

Halo with Parhelia (Hevelius).

Halos and the concomitant phenomena are more frequent in winter than in summer, and they are far more abundant the higher the latitude of the station. In the Arctic Regions they are very commonly observed, and

OPTICAL PHENOMENA. 205

the above are some sketches of halos which have been placed on record. It will be seen that the arrangements are very fantastic, and the variety of the manifestations very great.

Fig. 33 is a copy of a woodcut given by Loomis [1] of a halo belonging to a type not unfrequently seen. In the British Isles the phenomenon usually is more simple in its arrangement.

Fig. 34 represents a halo seen by Hevelius, Feb. 20, 1661.[2] The diagram gives a vertical projection of the sky.

As a rule, we may say that in these latitudes the appearance of a halo is a sign of unsettled weather, if not of a storm. It shows that a quantity of moist air is present in the upper strata of the atmosphere, and this has probably been produced by the outflow from an area of barometric depression (Chapter XVIII.) at no great distance, which as it advances will cause a storm.

Cloud colouring is also an optical phenomenon, depending on diffraction during the passage of the light through the atmosphere. Foreign particles suspended in the air, whether dust, smoke, or water droplets, obstruct the rays of light, offering greater obstacles to the passage of the rays of small than of those of great wave length—to the violet than to the red rays. These particles are for the most part accumulated in the lower strata of the atmosphere, and so, as the sun sinks, his rays have to pass successively through greater and greater thicknesses of these strata, and his light is, so to speak, sifted. The violet and blue rays are the first to disappear, so that orange

[1] Loomis' *Meteorology*, p. 216.   [2] *Mercurius in Sole Visus*, p. 173.

tints come out. The yellow rays are then stopped, and the last tints at sunset on the clouds are always red. Conversely, at sunrise the colouring gradually changes from red to clear white light, as the sun rises higher and higher.

It is also on account of impurities floating in the atmosphere, that in large towns the sun, both at rising and setting, usually appears red.

Another phenomenon, the explanation of which, like that of the rainbow, belongs to optical science, is that known under the name of 'Mirage.' In 'Nature,' for November 19 and 26, 1874, will be found an interesting lecture on these appearances by Professor J. D. Everett, F.R.S., from the opening paragraphs of which the following description is extracted:—

'The name of "Mirage" is applied to certain illusory appearances due to successive bending of the rays of light in their passage through the atmosphere. These appearances are by no means uniform.

'Sometimes, especially in hot countries, the observer loses sight of the ground beyond a certain distance from his position, and sees in its stead, what looks like a sheet of water, either calm or with movements, resembling waves; and if any distant objects are sufficiently lofty to be seen above this apparent lake, their images are seen beneath the objects themselves, inverted as if by reflection in this imaginary water. The dry and hot soil of Egypt is famous for the production of this form of the phenomenon. It is also mentioned as of frequent occurrence in the plains of Hungary, in the plain of La Crau in the south of France, and in the fen districts of England, when dried up by the summer heat. It is also common in Australia. The Deputy Surveyor-General of South Australia once reported the existence of a large inland lake, which on further examination turned out to be nothing but a mirage.

'Another class of appearances are known (especially among nautical men) under the name of *looming*. Distant objects are said to loom when they appear abnormally elevated above their true positions. This abnormal elevation not infrequently brings into view objects which under ordinary circumstances are beyond the horizon. It is also frequently accompanied by an appearance of abnormal proximity (though this may perhaps be rather a subjective inference from the unusual elevation and clear visibility of the objects, than a separate optical characteristic), and it is further accompanied in many, though not in all cases, by a vertical magnification, the heights of objects being many times magnified in comparison with their horizontal breadths, so as to produce an appearance resembling spires, pinnacles, columns, or basaltic cliffs. Some beautiful descriptions of these latter appearances, with illustrative plates, are given in Scoresby's " Greenland," the objects thus magnified being icebergs; and a very full and interesting account of the phenomena of mirage, as observed in high latitudes, will also be found in the " Arctic Regions " of the same author.

'It is usually across water that looming is observed; and as a surface of water stands naturally in contrast with a sandy desert, or a surface of parched land, so also the optical effects produced are, in a manner, opposite. The inverted images which are often presented in looming are not beneath the object, as in the case of mirage on dry land, but above it, as if formed by reflection in the sky. The only examples that I have myself seen of mirage were of this kind. They were seen across sheets of calm water, the hills on the other side being seen with fictitious hills upside down resting on the tops of the real hills. . . .

'There is always more or less of change observable in the images formed by mirage, and the changes are greatest and most sudden when the images are most distorted, as compared with the true forms of the objects. The appearances also change with the height of the observer's eye. Looming is seen to the greatest advantage from an elevated position, such as the masthead of a ship. The mirage of dry land is sometimes visible at any moderate heights, but in other cases—especially in countries which are not very hot—the range of height from which it is visible is extremely limited. Observers of mirage on the sands of Morecambe Bay, and of the Devonshire coast, state that it could frequently be only seen by stooping.

'Mirage is seldom seen in winter. The hot shining of the sun seems to be an invariable antecedent; and this is true even of the Polar Regions, where Captain Scoresby attributes the phenomenon to "the rapid evaporation which takes place in a hot sun from the surface of the sea, and the unequal density occasioned by partial condensations, when the moist air becomes chilled by passing over considerable surfaces of ice."'

Professor Everett goes on to give a very full explanation of the phenomenon, but it will suffice here to say that it is produced when the air is at rest, so that there is a horizontal arrangement of its successive strata from the ground upwards. If then, from any cause, the regular rate of decrease of temperature in a vertical direction is disturbed, as, for instance, by undue heating or cooling of the surface underneath, the rays of light, passing more or less obliquely, will suffer total internal reflection at the surface of the rarer stratum. They will consequently follow a curvilinear course as they pass through the successive strata, and be bent down or up towards the side on which the denser strata lie, and finally afford a picture of an object outside the direct line of view. What is seen over heated land is a portion of the sky depicted apparently on the ground; the denser strata are above, and the rays are bent upwards. Over the cold surface of the polar seas the denser strata are below, the rays are bent downwards, and objects below the horizon are thrown up.

All the varieties of the appearances presented by mirage, with distorted or inverted images, are capable of explanation on the principles indicated.

PART II.

CHAPTER XII.
THE DISTRIBUTION OF TEMPERATURE.

In Chapter II. it has already been explained that the sun is practically our sole source of heat; but in order to set forth the facts of the distribution of temperature in a clear light, it will be well to repeat some of the previous statements.

The sun's apparent motion round the earth in a day gives rise to a diurnal period of temperature. The sun has also an apparent motion in the ecliptic in the year, passing from one tropic to the other and back again, and this causes a corresponding annual period of temperature.

The effect produced by the solar rays depends in great measure on the angle at which they strike the surface of the earth. In the daily period the sun is highest, or most nearly vertical, at noon; and in the yearly period, outside the tropics, he is highest at the summer solstice,—June for the northern, December for the southern, hemisphere.

In the first place, it is evident that, if we consider the case of a cylindrical beam, its heating power will be concentrated on a smaller area the more nearly vertical its path is.

Secondly, a nearly vertical beam has to pass through a less thickness of the dense lower strata of the atmosphere than an inclined one, and we have already learnt from Chapter IV. that a large proportion of heat is absorbed by those strata.

The action of such heat as succeeds in passing through the atmosphere is not, in the first instance, to warm the air, but to raise the temperature of the surface of the earth, and the amount of rise of temperature produced by it depends in great measure on the nature of the surface on which the beams happen to fall, being greatest over bare earth—such as rocks or a desert, less over vegetation—such as forests, and least of all over the sea, for the reasons already explained in Chapter III., p. 46.

As soon as the heat from the sun has reached the earth, its warming effect on the air begins to be available; for it has now changed its character and has become 'dark' heat, that is, heat coming from a source of comparatively low temperature. Such heat is absorbed by the atmosphere with much greater facility than heat coming from the sun, as already explained in Chapter IV., p. 53.

When the earth is heated by the solar rays it begins to radiate to the atmosphere resting on it, and the amount of heat passing from the earth in this way is greatly affected by the condition of the sky, the radiation being much more active in the absence of clouds. As we have seen in Chapter VII., the presence of moisture in the air presents great obstacles to the escape of the heat radiated from the earth—a cloud covering reflecting it back to the earth almost entirely. The reader will remember that it has been explained that

THE DISTRIBUTION OF TEMPERATURE. 213

dew or hoarfrost cannot occur on cloudy nights, as the heat cannot escape from the earth into space.

The first action of the heat, when it has warmed the surface of the earth, is to warm the layers of the atmosphere in immediate contact with it, and as heat is very slowly communicated from stratum to stratum of the air, we find the lowest layers to be the warmest.

The reader need hardly be reminded that on mountain tops the temperature is, on the whole, much lower than on the plains below. This difference is far more marked in the daytime than at night, in summer than in winter. In fact, in the case of severe frosts the conditions are ordinarily reversed and the temperature rises with height instead of falling. This gives the well-known phenomenon of the 'up-bank thaw,' when it thaws on the hills while the frost is unbroken in the valleys below. I shall return to this subject in Chapter XVII., when speaking of climate.

Supposing that the air were perfectly dry, the rate at which the temperature would fall in an ideal vertical column standing upon the earth is 1° F. for every 180 feet; but as the atmosphere contains moisture, which is liable to be condensed by cold, giving out its latent heat in the process of condensation, the heat so rendered available will diminish the rate of cooling in the vertical column, and, as a very general rule, we may take about 1° F. for every 300 feet, the value given many years ago by Herschel, as the regular rate of decrease.

This consideration shows us that if we wish to lay down on a map the temperatures all over the globe, from the observations furnished by a number of stations situated at very various elevations above the sea, we must reduce the results to their equivalents at sea level, if we wish to attain anything like accordant figures, although the representation these afford has no real existence in nature.

Let us now return to the subject of the direct action of solar heat, and consider how it affects temperature, producing the daily and yearly march of that element.

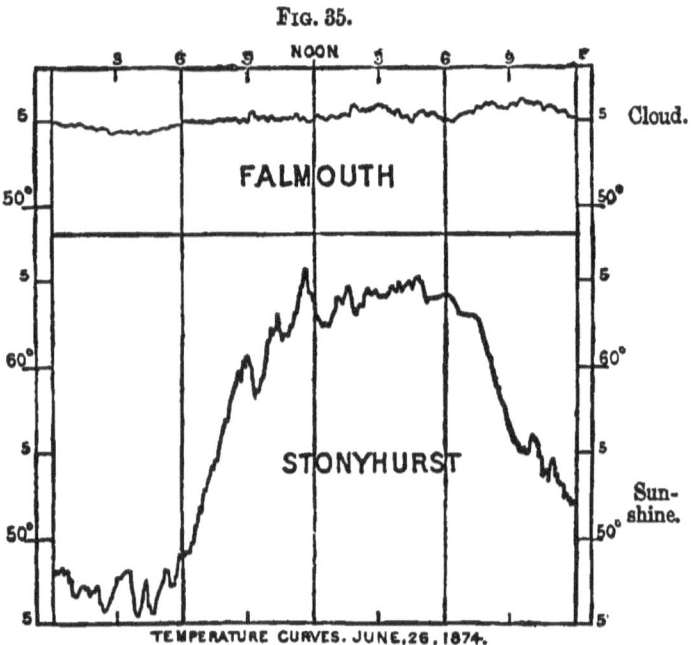

In Chapter III. we have treated in general terms of the movable equilibrium of heat, and have stated the reasons, based upon that principle, why the hottest portion of the day is not at noon, or of the year, in our latitudes, at midsummer, but we must now consider the subject more closely.

That the diurnal range of temperature is directly connected with the sun's rays is apparent from the fact that it is obliterated by cloudy weather or by fog, and is also entirely absent through the long and dreary winter of the Polar Regions. Figure 35 (p. 214) exhibits the continuous temperature curves for the same day at two stations distant not more than 200 miles apart. At Stonyhurst, where the day was fine and the sun shining, there was an extensive range of temperature of 20°, while at Falmouth, where there were heavy clouds and constant rain, there is hardly a trace of range during the whole day. The same fact comes out in other places. For instance, at Trevandrum, in the south of Hindostan, the diurnal range in January is about 17° and in July only 9°, a little more than half the former amount, although the sun is higher in the latter than in the former month. The reason is that in January the North-east Monsoon is blowing, with a clear sky, while in July the opposite wind, the Southwest Monsoon, prevails, with clouds and rain.

That the range is almost entirely due to the action of the heat of the sun's rays falling on the dry land is shown by the fact that at sea, close to the equator, it is hardly perceptible, while on a small island, like Ascension, it is well marked (fig. 36, p. 216). In both cases, however, the state of equilibrium, between loss and gain, is reached earlier than at a comparatively inland station such as Greenwich for example, and the apex of the curve approximates to noon.

The amount of this diurnal range therefore depends, in some measure, on the distance of the station from the sea-coast, for although we can see from fig. 36 (p. 216) that the curves at Greenwich, on the western coast of Europe, and at Barnaoul, in the centre of Siberia, have a close resemblance to each other, the latter is the more marked. Diurnal range is entirely absent in the Polar Regions, while the sun is below the horizon, as the curve for Boothia Felix shows. In the temperate zone it is greatest about the time of the equinoxes,

FIG. 36.

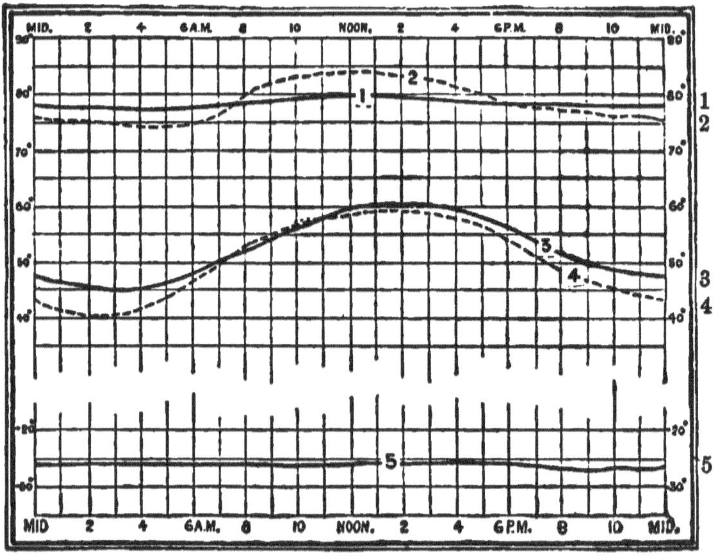

Diurnal Range of Temperature.

1. Equator at sea. May.
2. Ascension. "
3. Greenwich. "
4. Barnaoul. May.
5. Boothia Felix. January.

for then the days and nights are nearly equal in length.

The course of the curve of yearly temperature is governed by the same principles as the curve of daily temperature; the highest point does not coincide with the epoch of the greatest meridian altitude of the sun, but falls at a later period. Accordingly in these latitudes the hottest period is the end of July, as we have seen from the chrono-isothermal diagram for Greenwich (p. 48); but the actual form of the annual curve, like that of the diurnal, though mainly regulated by latitude, depends also on other conditions.

In the torrid zone, between the tropics, the sun, as has already been said (p. 9), is vertical over every spot twice a year. On the equator itself, these two periods of vertical sun occur six months apart, at the equinoxes, and the hottest periods, therefore, ought to be in spring and autumn; but as the period of the greatest altitude of the sun may coincide with that of heavy rains, as has just been stated (p. 215), it is clear that the highest point of the curve is not fixed solely by the sun's position.

The nearer we approach to either tropic the closer do these periods of vertical sun come to each other, as the sun is advancing to the tropic and receding from it again. Within the torrid zone, therefore, the course of the annual curve is not so simple as in higher latitudes, while that of the diurnal curve shows comparatively little change throughout the year.

The diagram (fig. 37, p. 218) exhibits this peculiarity of the tropical curves very well. It shows the annual range and the mean temperature for four stations, two of them (Ascension and Calcutta) in the torrid, the other two (Greenwich and Barnaoul) in the temperate zone.

The range curve for Ascension (No. 1), and in a much greater degree that for Calcutta (No. 2), follow an irregular course, compared with those of Nos. 3 and 4.

In the temperate zone, the length of the day varies as the latitude increases, from the tropics to the polar circles. In these zones, therefore, the annual curve is fairly regular, but the course of the diurnal curve is largely modified from month to month according to latitude, for at the respective summer solstices on the

218   ELEMENTARY METEOROLOGY.

Arctic or Antarctic Circle the day is of course twenty-four hours long.

In the frigid zones there is at one season no day, and at the opposite season no night, and at the poles themselves there is of course, theoretically speaking, one day and one night in the year, each of six months'

FIG. 37.

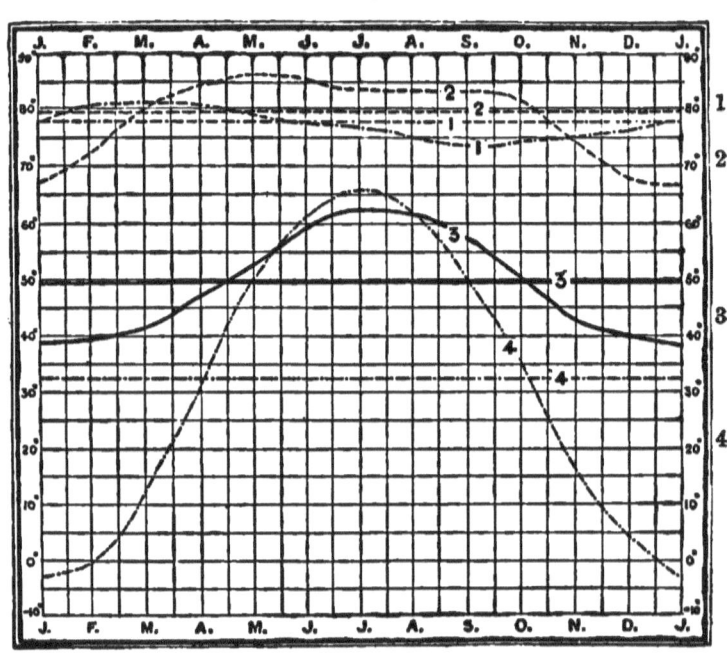

Mean Temperature and Annual Range Curves.

1. Ascension. 3. Greenwich.
2. Calcutta.   4. Barnaoul.

duration. In these regions, therefore, the effect of the sun's heat is exerted in a totally different way from that which obtains in lower latitudes, and the annual curve is as much distorted as the diurnal.

Both classes of range curves, the diurnal as well as

THE DISTRIBUTION OF TEMPERATURE. 219

the annual, are greatly modified according to the situation of the station, whether it be on the sea-coast or in the interior, and according to its elevation above the sea-level, as we shall see when we come to treat of the distribution of land and water on the globe and of the climate of elevated districts and mountains. (Chapter XVII.)

As regards the action of land and water, the surface of the earth presents us with three conditions :—

(*A*). Constantly solid (the continents).

(*B*) Constantly liquid (the sea in low latitudes).

(*C*) Sometimes liquid and sometimes solid (the Arctic and Antarctic Oceans).

The actual distribution of land and water on the globe is very irregular, as we have seen in Chapter II. In the first place, the proportion of land to water is only about 1 to 4; and in the second, most of the land is collected in the northern hemisphere, in the proportion of 13 to 1. In fact, if we take a hemisphere, having its centre in the British Isles, we shall find that it contains almost all the land. Only $\frac{1}{27}$th part of the existing land has land at its antipodes.

There are certain broad principles to be taken into consideration in explaining the effect of this distribution of land and water on climate or on the course of the lines drawn on *isothermal* charts, *i.e.* charts intended to represent the temperature of the globe.

The first principle is that as the specific heat of water is five times that of dry land (earth or rocks), it takes five times as much heat to raise a given mass of water through a given range of temperature as it does to raise an equal mass of dry land.

The second principle is that the absorbing and

radiating power of water is far less than that of dry land.

In other words, we may say that it is very difficult to heat up the sea-surface, but that when once it is heated it retains its temperature and does not materially warm the air lying above it, while the contrary is the case with dry land.

A third principle is that in those parts of the globe where the sea is frozen during any part of the year we meet with an important agency tending to moderate climate and smooth down its extremes.

In Chapter VII. the latent heat of steam has been mentioned, and its effect in raising the temperature of the air when the steam is condensed to the form of cloud. The latent heat of water also plays an important part in climatology. When in winter the sea freezes, a quantity of latent heat is set free which tends to moderate the extreme rigour of the cold and protract the autumn, but when the thaw comes at the end of winter and the ice breaks up, the demand for heat from the air to effect the change of state of the water exerts a chilling influence and delays for a considerable time the coming of spring weather.

As a deduction from these principles we see the truth of the following statement.

The presence of land close to the equator raises, and its presence near either pole lowers, the mean temperature, and, conversely, equatorial oceans lower, and polar oceans raise, the mean temperature.

Sir Charles Lyell, in his 'Principles of Geology,' endeavours to show how the successive variations which are proved, by geological evidence, to have occurred in the climate of the earth, might have been produced, without any disturbance of the relative proportion of land to water, by simply shifting the continents; an arrangement by which all the land should be situated in a belt extending 30° on each side of the equator, giving the greatest possible amount of heat, and a converse arrangement, by which no land should extend to within 40° of the equator, giving the greatest possible amount of cold. Such a hypothesis as this is, of course, purely imaginary, but it will serve to fix the reader's ideas; and, that the assumed cause is not inadequate to produce the effects attributed to it will appear probable when we come to consider the temperature charts which will shortly be described.

Heat is transmitted with great facility through liquids and gases by means of convection currents, and it is mainly the currents of the ocean and the atmosphere which moderate the climates of the regions over which they prevail. In fact, as Mr. Croll has said, 'without ocean currents the globe would not be habitable.'

It is somewhat anticipating what I shall hereafter have to say about currents to point out that the Gulf Stream issuing from the Gulf of Mexico, and flowing in a broad current into the Atlantic, makes its influence felt even within the Arctic Circle by keeping the harbour of Hammerfest, in 71° N., open in the depth of winter, while the opposite shores of Labrador, more than twenty degrees farther south, under the influence of the American Arctic current, are icebound for the greater part of the year.

The Pacific, too, has its warm current—the Kurosiwo, or 'Black Stream' of the coast of Japan—which is estimated at three times the volume of the Gulf

Stream, and which confers on the shores of Alaska and British Columbia their freedom from ice in winter, but of which the influence does not reach to the same latitude as that of the Gulf Stream on the coast of Norway, owing to the configuration of the land to the northward of the Pacific Ocean.

Another most influential agent in conveying heat from warmer to colder regions is the wind, and it is our prevailing West winds which, bringing, as they do, warmth and moisture from the sea in winter, carry on the benefits of proximity to the ocean to the internal portions of the British Isles.

Let us now consider the distribution of temperature over the globe, as shown on the Charts—Plates I.–III.

The idea of representing this distribution by means of curves, or so-called isothermal lines, is due to A. von Humboldt, and was worked out by Dove, whose charts were published by the British Association in 1853, and have since been somewhat modified by Buchan and others, by the use of the material which has accumulated during the last thirty years. Of course, in order to give a fair representation of the changes throughout the year, there should be a chart for every month; but we must content ourselves with three charts—one for the year and one each for January and July, as the months which are generally warmest and coldest respectively.

The representations on these charts have been copied from those given by Professor Mohn in his 'Grundzüge der Meteorologie,' 2nd edition (Berlin: D. Reimer, 1879). As regards the Russian Empire, however, the curves given by Professor Wild, in the Atlas to his great work, 'Die Temperatur-Verhältnisse des russischen Reiches' (St. Petersburg, 1881), have been followed. In both cases, of course, such changes have been made as were required by the conversion of the Centigrade to the Fahrenheit scale.

The curves on these charts are called Isothermals (curves of equal heat), and on Plate I. they are drawn for every 10° of mean annual temperature. The curves on Plate II., for January, show the winter temperature of the northern hemisphere, and those on Plate III., for July, its summer temperature.

Plate I. shows us the mean annual temperature, the isothermals in blue being those for 32° and under, the rest being in red.

The first point that strikes us is that the curves over the northern hemisphere are far more irregular in their course in high latitudes than those in the southern. This arises from the uneven distribution of land and water which has already been explained, and the relatively great preponderance of the former north of the equator. A second remarkable feature is the great development of blue lines near the North Pole. This is, however, in part owing to the fact that, while the map extends beyond 80° N., it stops short at 70° S., and the projection adopted, that of Mercator, exaggerates the area of polar countries enormously; so that the North Polar Regions are very prominent, while the corresponding district near the South Pole is not seen at all. But still, as we shall subsequently see, the northern hemisphere, in high latitudes, is really colder than the southern. In fact, on the parallel of 60° we find a mean temperature below 20° near Hudson's Bay and in Eastern Siberia, while in the corresponding southern latitude the mean temperature is at least 10° higher.

To come to the actual isothermal lines, the highest on Plate I. is that for 80°, which surrounds two spaces —one covering the north of South America, and the other extending northwards from the Guinea coast to near Suez and southwards to the Seychelles, thence stretching in a broad belt of nearly uniform width to the meridian of 100° E., and gradually thinning down to its disappearance near the equator in 140° W. These patches—for they do not form a continuous belt —are called, by a misnomer, the thermal equator. It will be seen that most of its course lies to the north of the terrestrial equator, owing to the amount of land in northern latitudes.

The next curves, or pair of curves, are those for 70°, and both of them trend towards the equator on the western coasts of the Americas, and turn very suddenly in the opposite direction as they enter on the continent. The same relation, but in a less degree, appears on the coasts of the old continent, but the change is not so sudden. The northern curve approaches Tunis, skirts the southern shore of the Mediterranean, and attains its highest latitude in Central Asia. The curve of 60° in the southern hemisphere almost follows the parallel of 35° S., the amount of land in those latitudes being too small to produce any material irregularity. For the same reason the next isotherm, that of 50°, almost coincides with a small circle of the globe in lat. 42° S.

In the northern hemisphere the isotherm of 60° does not show any very striking sinuosities, and that for 50° resembles it in this particular, but of course lies in higher latitudes. However, on the western coasts of Europe and America, as compared with the eastern, its position exhibits the marked difference in latitude which is so strongly characteristic of the curves which remain to be described.

The line for 40° enters the American coast in lat. 58°, and leaves it at the mouth of the St. Lawrence in about 48°, while over the Old Continent its course is even more sinuous. It reaches 68° N. near the south point of the Loffoden Islands, then passes nearly due south for about six degrees, trends south-eastwards, and finally leaves the coast of Asia, at the south point of Saghalien, in latitude 45°.

We must now turn to the blue lines. The curve for 32° touches the parallel of 50° at the south end of Hudson's Bay, and passes below it on the Amur in 120° E., while it reaches to the northward of the North Cape in lat. 72° N. It takes two remarkably sharp turns: one in Lapland, where it actually runs back, south-westwards, to the head of the Gulf of Bothnia; the other in Siberia, between 100° and 110° E., where it takes a bend round Lake Baikal; this sheet of water, from its enormous depth (over 2,000 fathoms), being very slow to be affected by changes of air temperature, and, accordingly, leaving its mark on every isotherm throughout the year.

The position of the corresponding line, for the freezing point, in the southern hemisphere is almost purely hypothetical, and the same must be admitted of great portions of the courses of some of the lower isotherms in the northern hemisphere, for it is only on the west coast of Greenland and in Siberia that any regular observations are available to fix their positions.

On the Old Continent the lowest isotherm, that for

0° F., forms a closed curve near Jakutsk, in about 135° E., but over the region lying within the Arctic Circle, off the north coast of America, portions of two curves are shown for 0° and −10° F.

The annual isotherms, however, do not convey any correct impression of the real conditions of climate of the globe, for, as already stated in Chapter III., the bare statement of an average value gives no idea of the extent of variation among the actual figures from which that average has been derived. To take an example. Dublin and Astrachan differ only 1°·1 in their mean annual temperature, that of the former being 50°·2 and that of the latter 49°·1; but while the temperature of January at Dublin is 41°·2, at Astrachan it is 20°·5, being a difference of 20°·7. In July the contrast is nearly as great, for Dublin has a temperature of 60°·4, being 15°·7 colder than Astrachan (76°·1). It is needless to multiply instances; they will be evident from inspection of the maps for January and July, and of the Range Map, Plate XI. (Chapter XVII.).

The second map we have to consider is that for January (Plate II.), when the curves have their greatest southerly displacement. The isotherm of 80° approaches the parallel of 30° in South America, in South Africa, and, as we may assume, in Australia. The other southern isotherms run nearly along the parallels, excepting that they all rise in latitude about the meridian of Cape Horn and sink again before they reach that of the Cape of Good Hope.

In the northern hemisphere, however, we find the most remarkable contrasts in the thermal conditions. In the first place, the northern isotherm of 80° does not continue in north latitude throughout its course,

THE DISTRIBUTION OF TEMPERATURE. 227

but remains below the equator from 110° E. to 120° W. The curves for 70°, 60°, 50°, and 40° all exhibit a bend polewards on the western coasts of both America and Europe, and this feature is more marked the lower the temperature, the curve of 40° having a range in latitude of over twenty degrees, from 52° N. in Ireland to 31° N. in Mantchuria. The isotherms for lower temperatures, however, exhibit much more irregular courses, that for 32° forming a loop which reaches as far as 72° N., off the Scandinavian peninsula, trending somewhat westwards just inside the coast of Norway and passing nearly due south to the north of Italy. It dips farther and farther south as it crosses Asia and touches the parallel of 32° N. in Western China. This line, therefore, has a range of forty degrees of latitude.

The isotherms of 20° and 10° are shown by Prof. Wild to take bends, even much sharper than that just described, over Lapland and Northern Finland; they then pass due south-eastwards to the Caspian, remaining nearly parallel to each other, and rising respectively to lats. 60° and 62° in the Territory of Alaska. They dip down to cross Lake Superior, and rise in a due north-east direction to the points where we commenced to trace them. It is evident that much of the course of these curves, over North-west America and the Arctic Ocean, is hypothetical, from the deficiency of observations.

The curve for 0° F. passes near York Factory, at the southern end of Hudson's Bay; it then sweeps across Labrador to Disco, across Greenland to Spitzbergen, and thence to the west coast of Nova Zembla, whence it dips due southwards, and its course over Siberia has been well fixed by the Russian observations.

It exhibits a peculiar bend eastwards, and then southwards again, about 62° E., sweeping round the catchment basin of the Irtysch—a district that produces a distortion of the isotherms in several months. The line subsequently approaches very near the head of the Gulf of Pechili, between China and the Corea, and finally leaves the Asiatic coast near Petropawlowsk, in Kamtschatka.

The line for −10° is shown in separate portions. The Asiatic part shows the warming influence of Lake Baikal as it bends round the northern and eastern sides of that immensely deep sheet of water, whereas the annual isotherm, as we have seen (p. 225), and as we shall see, that for July also, are made to dip southwards, passing round the western and southern shores of the lake.

On the American side the curves below 0°, as far as −40°, are not drawn over Greenland, or the land lying to the west of it, a part of that for −40° being shown to the north of Melville Island, and touching Sir E. Belcher's winter station at the head of Northumberland Sound. In Asia, however, the line for −40° is well defined, and forms a closed curve passing round Jakutsk and the mouths of the Lena, while inside it a small curve of −50° surrounds Werchojansk, the coldest inhabited place in that month, Ustjansk being just outside the line.

Turning now to Plate III., for July, we find the conditions in general much more equable, the blue lines having almost entirely disappeared.

Commencing at the south, the curve for 32° is shown off Cape Horn, about the parallel of 60° S., and it is drawn round the globe between that latitude and the parallel of 50°, although there are hardly any observations to fix its position in such inaccessible regions.

The line for 40° passes near Cape Horn, and is met with off the Cape of Good Hope in 48° 30′, but passes to the southward of lat. 50° S. for a short space—about the meridian of 180°.

The curves for 60° and 70° show the influence of cold Antarctic currents on the western coasts of both continents. That for 80° dips to about 14° S. in South America and to 11° S. in Africa, to the southward of the central lakes. It lies on the northern side of the equator in all the oceans.

Over the whole extent of Northern Central Africa, Arabia, and Persia, up to the Punjaub, an area with a temperature of 90° is shown. This is the region of highest mean monthly temperature on the earth.

The northern curve of 80° turns sharply northwards through Mexico into the Western States beyond 40° N.; it then runs south-eastwards below the parallel of 20° in mid-Atlantic, and rises to skirt the north coast of Africa; thence its way lies, through the Caspian and the Desert of the Great Gobi, to the North of China, again descending nearly to 10° N. in the Central Pacific.

The line for 70° follows a very sinuous course. It runs northward and westward through California into British America, forming a loop which reaches the parallel of 58° N. in about 122° W. Returning southwards it leaves the coast near Portland, in Maine, and dips southwards almost to Bermuda. It then gradually rises, crossing the Old Continent, till it betrays the influence of Lake Baikal by a very sharp loop, after which if again takes a sudden turn southwards along the coast as far as to the Corea.

The line for 60°, lying near London, trends due north-east to the head of the Gulf of Finland, and maintains the latitude of the Arctic Circle almost up to 140° E., when it dips southwards more than fifteen degrees to pass round the Sea of Ochotsk.

The lowest isotherm on the chart, that for 40°, passes to the north of America, is bent northwards in Davis's Straits, descends again, and rises to touch Spitzbergen. It then passes between Nova Zembla and the mainland, and sweeps through the Tundras of North Siberia to the New Siberia Group, again descending nearly to Behring's Straits.

When we compare again the curves on Plates II. and III. we notice how slight is the seasonal displacement of the isotherms in the southern hemisphere as contrasted with that found in the northern. Accordingly, when the northern hemisphere is in winter and exposed to severe cold, the southern has generally a cool summer; while, conversely, during the intense heat of our summer the southern latitudes enjoy a comparatively warm winter. We are thus led to expect that the general temperature of the earth in July must be higher than it is in January, and there is some evidence that this is the case; for Dove has calculated that the temperature of the globe, as a whole, is 62°·6 in the former, and only 54°·5 in the latter month, so that there is a difference of about 8° in favour of July; the reason being, that in the summer of the northern hemisphere the sun's rays fall on the greatest possible land area.

Dove has also calculated the mean temperatures of

the two hemispheres, and finds that of the northern to be 59°·9 and that of the southern 56°·5, thus affording another proof that the great water hemisphere is the cooler.

These results are very remarkable. The reader will remember that, as already stated in Chapter II., the orbit of the earth is elliptical, the sun being in one of the foci, and that the time when we are farthest from the sun corresponds with the northern summer; and the time when we are nearest to the sun, with the northern winter. Accordingly the earth is really nearer to the sun in the southern summer than in ours, but a compensation to a slight extent is found for this differ ence in the fact that the northern summer is three days longer than the southern (p. 10), so that the increased effect of the sun's rays is exerted for a shorter time.

Nevertheless, we should anticipate the southern hemisphere to have hot summers and cold winters, while, as we have just seen, the opposite is the case, owing to the unequal distribution of land and water.

In order to show the great contrasts which exist between climates on the same parallels of latitude, Dove, in his great work before mentioned, calculated, for each month, the normal temperature of every tenth parallel of latitude, *i.e.*, that which it would have everywhere if its actual, but variously distributed, temperature were uniformly arranged, and gave values for the mean of the year, which have been corrected by Hann[1]:—

Lat.	0°	10°	20°	30°	40°	50°	60°	70°	80°	90°
Northern	79°·7	70°·9	77°·5	69°·8	56°·5	42°·4	29°·8	16°·0	6°·8	2°·3
Southern	79°·8	78°·0	74°·1	66°·9	55°·4	43°·7	32°·5			

---

[1] *Sitzungsberichte der Academie der Wissenschaften*, Wien, Jan. 1882.

No materials were available for Dove for the calculation of the values for the higher southern latitudes; in fact, those given for 50° and 60° S. have been determined by Dr. Hann from data collected of late years. Dove drew maps of what he termed *the thermic anomaly* of each month, representing this by curves called 'Isabnormals.' Those for January and July, as being the most interesting, are reproduced in Plates IV. and V. These curves represent equal differences from the normal temperature, and where these differences are positive, where the climate is too warm, the lines are red; where the differences are negative, indicating cold, the lines are blue. The line connecting places which are neither too hot nor too cold is drawn double, red and blue.

In January, Plate IV., the whole of Asia is coloured blue, and around Werchojansk we have a closed curve, indicating 40° below the normal, while in slightly higher latitude on the Arctic Circle, and on the meridian of Greenwich, we find a favoured region, as many degrees too warm. The successive isabnormals run very close to each other along the east coast of Asia, and the normal line passes nearly parallel to the coast from Behring's Straits to the western end of New Guinea.

North America also shows negative anomaly, but only to half the intensity of the Siberian cold region.

The district of greatest positive anomaly is the North Atlantic, and this feature increases with latitude to its maximum of $+40°$, already mentioned, on the parallel of the North Cape.

The Northern Pacific, like the Atlantic, is marked by red lines, but the contrast between its relatively

THE DISTRIBUTION OF TEMPERATURE. 233

hottest region and the coldest in Arctic America is only one half as great as that noticed between Siberia and the North Atlantic.

In the southern hemisphere, the Southern Indian Ocean, and the Atlantic, up to 20° W. are too cold, and the same conditions hold over most of Africa, even reaching to the deserts of the north and to Egypt; but the anomaly is nowhere great, its highest value being 10°, off the Cape of Good Hope. Another large cool area stretches in a peculiar shape over most of the Southern Pacific, but no part of it is even 10° too cold.

In July (Plate V.), as might have been expected from the appearance of the isothermal lines, the thermal disturbance is not nearly as great as in January, no part of the globe showing a greater deviation from the normal than about one fourth of that which obtains in the latter month. In the summer of the northern hemisphere the continents show positive anomaly, but to the maximum extent of only 10°, over two areas, one stretching from the desert of the Great Gobi across Persia and Arabia, the other in the very region characterised by intense excess of winter cold in Siberia.

The oceans in the northern hemisphere are both within the blue areas, and in high latitudes show the cooling effect of melting ice, for in Davis's Straits the negative anomaly is 10°, and where the cold water pours into the Pacific through Behring's Straits it rises to 15°.

In the southern hemisphere a cold area is shown off the coast of South America in 20° S., caused by the chilling influence of Humboldt's current, and another of similar intensity appears over the southern part of Australia in the middle of the southern winter.

The Indian Ocean and most of the South Pacific

are too hot; one district of 5° of positive anomaly, off Natal, being distinctly due to a current of warm water flowing down the Mozambique Channel (p. 308).

It may be interesting, before leaving the subject of the isothermal charts, to say a few words about the extremes of temperature which have been experienced. It was formerly the fashion to look for poles in connection with all terrestrial phenomena, and Sir D. Brewster, in particular, maintained that there must be three poles of cold—two in the northern, and one in the southern hemisphere. Recent observations have, however, proved that this idea was not quite correct, being founded, like many other generalisations, on insufficient data.

Although at Werchojansk the temperature of January is lower than anywhere else, where continuous records have been kept, the weather in July in that part of the world is very fairly warm, the mean being within one degree of that of Dublin, while at Rensselaer Harbour (Dr. Kane's winter station), and in other stations in Smith's Sound, &c., the temperature of July hardly rises above the freezing-point. The highest temperature observed at Rensselaer Harbour was 55°, while at the Siberian winter pole of cold, the summer heat has several times reached 96°.

Dove gave a table of the lowest mean monthly temperatures on record, which is now somewhat out of date, having been published in 1855, but it clearly showed that it was only in December and January that the Siberian stations enjoyed the unenviable position of being the coldest inhabited places; throughout the remainder of the year the American Arctic Regions far outdid them in that particular.

The highest monthly average known is that of May for Massowah, which amounts to 99°, and the difference between this and the lowest for January, −56° at Werchojansk, is 155°. The contrast between individual temperatures actually observed is, however, far greater than this. Instances of the temperature to which a sandy soil may rise have been given at p. 46. As to instrumental observations of air-temperature, 130° has been recorded at Mursuk in Fezzan; and at Cooper's Creek in Australia, where Burke and Wills died, a thermometer graduated up to 127°, and left for safety in the fork of a tree, was found to have been burst by the expansion of the mercury. On the other hand, Gorochow recorded −81° December 30, 1871, at Werchojansk, and Kane noticed −69° at Rensselaer Harbour.

Accordingly we see that the human system can bear a difference in the air-temperature of over 210°, or of thirty degrees more than the interval between melting ice and boiling water. As regards the relative endurability of these extremes, Dr. Moss, in his 'Shores of the Polar Sea,' has some remarks to the point. He says:—

'Many a time the relative merits of Arctic cold and Tropical heat were warmly canvassed. Many of our officers and men had lately returned from the Ashantee campaign, and they could speak with authority. There was one thing clear, one could sometimes get warm in the Arctic, but never get cool on the Coast.'

So much for temperature at or near sea-level. I shall return to the question of climate in a later chapter, but it will be interesting to say a word or two about the annual march of temperature on mountain-tops, as far as the observations available enable us to exhibit it. It has been already stated that, as a rule, temperature decreases with height. During the year 1865-6 observations were made on the Col St. Théodule, near Zermatt, about 11,000 feet above the sea, and of late years, at a height of over 14,000 feet, on Pike's Peak, Colorado, by the Chief Signal Office of the United States.

The extreme of cold registered on the Col St. Théodule was only 6°·5 F., so that there was no severe frost at all, but the temperature was uniformly low, for the thermometer never rose above 32° from November to April. The results for the summer temperature were, however, the most remarkable, the average of the three summer months being 32°·3, and that of July 33·°5. These are the lowest summer temperatures as yet recorded at any station, and they teach us how very greatly the amount of range is influenced by elevation.

Dr. Hann[1] has discussed these results, and has calculated from the general features of temperature at high levels in Europe a table for its diminution with height. This table gives, for the Western Alps, 273 feet as the difference of level corresponding to 1° F. of reduction, in July. Consequently, assuming the mean temperature at sea-level for Switzerland to be about 73° in that month, the level at which the mean July temperature of Northumberland Sound (36°·7), about the lowest we know of, would be met with in the Alps, would be 10,000 feet, while the average annual temperature of −2°·5, that of Rensselaer Harbour on Smith's Sound, would not be found until the level of

[1] *Zeitschrift der Oesterreichischen Gesellschaft für Meteorologie*, vol. v.

about 18,000 feet, *i.e.*, more than 2,000 feet above the summit of Mont Blanc.

The July temperature of Pike's Peak is 42° F., but then it must be remembered that this mountain is seven degrees nearer the equator than the Col St. Théodule.

CHAPTER XIII.

THE DISTRIBUTION OF ATMOSPHERIC PRESSURE, AND
THE CIRCULATION OF THE ATMOSPHERE.

THE distribution of atmospheric pressure over the globe, although, as we shall see, it is closely connected with that of temperature, did not attract much notice from scientific men for nearly half a century after Von Humboldt had introduced the drawing of isothermal lines. The credit of first publishing isobaric charts, and connecting the winds with variations of pressure, belongs to Buchan, whose paper on the subject appeared in 1869,[1] and whose curves have hardly been altered materially by subsequent investigators.

Later on we shall find how close is the relation of pressure distribution to that of temperature, for it is impossible to deal with the former without constant reference to the latter. I shall therefore commence by pointing out the broad principles of the action of temperature on pressure, and of the great atmospheric circulation thereby set in action, as explained by Hann.

If we were to suppose the earth at rest, with its entire surface at a uniform temperature, the atmosphere would form an envelope consisting of layers,

[1] 'The Mean Pressure of the Atmosphere and the Prevailing Winds over the Globe for the Months and for the Year.' *Transactions of the Royal Society of Edinburgh*, part 2, vol. xxv. Edinburgh, 1869.

or shells, concentric with the globe; the pressure being the same throughout any one layer, but varying from layer to layer according to a law depending on the distance of the layer from the earth's surface, or on its altitude. There would be no tendency to any disturbance of the equilibrium—there would be no motion of the air—no wind.

If now any portion of the earth's surface be heated, the air resting upon it will be expanded, and the equilibrium will be disturbed.

The action of this heating will be to cause the layers, or surfaces of equal pressure, to expand like the outside of an inflated bladder, and therefore to rise from the earth. The lowest will be most raised, and the pressure at the upper levels will increase.

It is easy to show how this must be the result. If we imagine a column of air confined in a chimney, extending above the limit of the atmosphere, and therefore only able to expand in a vertical direction, and if we apply heat from below the total weight will be unchanged, as no air can escape from the chimney laterally, while at any section taken above the base, a greater amount of air will be above that level when the column is heated than when it is at its ordinary temperature, and the pressure at such upper section will be increased. We can prove that this action really goes on in the atmosphere by the following comparison of the monthly mean readings at different levels in winter and summer:—

Place	Elevation, Feet	Pressure January, Inches	Pressure July, Inches	Difference, Inches
Geneva	1,335	28·66	28·66	0·00
Great St. Bernard	8,174	22·11	22·39	0·28
Col St. Théodule	10,899	19·77	20·16	0·39

This shows us that while at the lowest of these stations the pressure was the same in January as in July, the increase from the former to the latter month was augmented with the altitude of the station, and at the level of 11,000 feet was as much as 0·39 inches, nearly one-fiftieth of its total amount.

If we apply these principles to the atmosphere, as a whole, remembering that the mean temperature, in general terms, decreases from the equator to the poles. we shall see that the tendency will be for the surfaces of equal pressure, supposed in the first instance to be spherical shells, to assume the form of ellipsoids with their major axes perpendicular to that of the earth. The atmosphere will stand highest over the equatorial regions, as the district where the air is most expanded by the action of heat, &c., and accordingly the pressure in the upper levels in low latitudes will be greater than at an equal elevation above the earth's surface in high latitudes.

The result must be that the air at the upper level near the equator receives a tendency to flow off towards the poles, on either side, in order to produce an equality of pressure at an equal distance above the earth's surface all over the globe.

Accordingly, the heat sets the upper strata in motion, and the efflux thus produced first reduces the weight of the atmosphere near the equator, causing the barometer there to fall, and then conversely causes the barometer in higher latitudes to rise, as the influx of air from the equatorial regions adds to the pressure of the region at which it arrives.

That this is actually the case is shown by the difference between the pressures at different levels at the equator and near the fortieth parallel, from the observations on Antisana in Ecuador, and on Pike's Peak in Colorado. The means are for our winter (December to February) :—

	Equator	Lat. 39° N.
Pressure at sea-level	29·88 inches	30·20 inches
„ 13,000 feet	18·55 „	18·04 „

The pressure therefore in the lower stratum increases with the latitude, while in the upper the reverse condition holds, and the pressure at the equator is the greater.

This state of affairs must give rise to two currents (in each hemisphere separately)—an upper one from the equator towards higher latitudes, and an under one from higher latitudes towards the equator.

As the form of the earth is spherical, the circumference of the equator, a great circle, exceeds that of the smaller circles of latitude, which gradually diminish to a point at the pole. The meridians therefore converge as the latitude increases, and the air, flowing polewards, is moving in a bed continually becoming narrower and narrower, and is finally forced down to the surface of the earth long before the pole is reached.

Observations show that practically these upper currents descend to the level of the sea at about the thirtieth parallel of latitude, producing, by their mechanical action, the calms known respectively as the calms of Cancer and Capricorn, and giving rise to the high pressures which are found to exist at about the distance of thirty degrees from the equator in either hemisphere, wherever the circulation is not disturbed by irregular local heating of the earth's surface underneath.

This principle, as it attributes the primary movement of the air to the summation of the expansive action of the air layers throughout their entire depth, makes this movement in each column of air to depend on the temperature of the column as a whole, and not only on that of its lowest stratum which is immediately in contact with the earth, and is therefore most expanded, over localities where the temperature, as shown by the Isothermal Charts (Plates II. and III.), is from time to time the highest.

There may be locally overheated areas, producing local circulations in the lower strata—the Land and Sea Breezes, the West Monsoons of the Line—and even the great Monsoon system of the Eastern Hemisphere, but the general movement of the atmosphere proceeds uninfluenced by such local accidents as diurnal or even annual range, nay more, by the character of the earth's surface, whether it be water over the oceans or dry land over the continents.

In fact, the surface-heating is really more regular and effective in the case of the oceans than in the case of the continents, for, although the sea-surface is not so much heated by day, it is hardly cooled at all by radiation at night, and so it preserves a comparatively equable temperature throughout the twenty-four hours, the water making up by constancy of action what it wants in intensity.

The overflow of air in the equatorial regions takes place at a great height, for the level of the countercurrent has never been reached by the ascent of any of the peaks in the equatorial Andes; but that the air does flow away in the upper strata is abundantly proved by the fact that clouds are often noticed moving in a direction contrary to that of the Trade wind below.

A proof even more striking is afforded by the fall of ashes from the volcanoes in Central America in the Spanish Main, and from those in the Dutch East Indies in the Western Pacific.

An instance of the former occurrence was recorded in January 1835, when ashes from Coseguina, on the Bay of Fonseca, were borne in four days to Kingston in Jamaica, in the teeth of the Trade wind. The latter phenomenon took place in 1815, when the ashes from Tomboro in Sumbawa were carried to Amboyna, a distance of 800 miles, athwart the direction of the South-east Trade.

When we reach the outer edge of the Trade-wind zone we find the counter-current gradually descending to the sea-level. On the Peak of Teneriffe (lat. 28° N.), and on Mouna Loa in the Sandwich Isles (lat. 20° N.), there are constant South-west winds at the summit, and the former locality offers a very instructive illustration of the annual changes in atmospherical circulation. As the autumn comes on the sun moves southward, the Trade winds follow him, and the South-west wind creeps slowly, day by day, down the mountain side, until in winter the whole of the island, from summit to base, is enveloped in this return current, which in summer only touches the highest peaks.

When these upper currents reach the earth, about the latitude of the tropics, a great portion of the air which goes to form them is caught in the under-current and drawn back towards the equator, but the remainder flows on; and outside the parallels named in both hemispheres we can no longer trace the persistence of upper and under currents at all seasons, but the predominant direction of the air-motion is from the equator towards the pole, and this appears to increase in steadiness and force with height above the ground. The lower currents in high latitudes are chiefly regulated by the contrasts in temperature between the continents and oceans, and by the differences in pressure induced thereby.

The under-currents in low latitudes are those which first attracted attention owing to their constancy, and they were long held to afford the key for the interpretation of all atmospherical circulation. These are the famous Trade Winds, on either side of the equator, blowing so continuously in one direction that Varenius tells us that the sailors leaving Acapulco for the Philippine Islands might lash the helm and go to sleep for the whole run, as they could not fail to make their port; and the Spaniards called the Trade-wind region *El Golfo de las Damas*, for when once it was reached a girl might take the helm.

These winds blow towards the thermal equator, and the first attempt at a correct explanation of their movement from N.E. and S.E., instead of from N. and S. respectively, was given in the 'Philosophical Transactions' for 1735, by George Hadley.

The principle which he employed was that of the unequal velocities with which particles on the different circles of latitude are carried by the diurnal rotation of the earth, as explained in Chapter II. This would cause winds flowing from low latitudes into high to become gradually Westerly; and *vice versâ*, winds flowing from high latitudes into low to become gradually Easterly.

This explanation, however, though it was adopted and developed by Dove and so has become widely known, only applies to the North and South components of the wind's motion. Foucault's experiments with the pendulum, in 1850, showed that the deflection had nothing to do with the direction of the motion, but was uniform with winds from all points of the compass, depending simply on the latitude and the velocity of the wind. If ω be the angular velocity of the earth's rotation, v the velocity of the moving particle (in this case the wind), and λ the latitude, the deflection per second is $v \omega \sin \lambda$. This principle, therefore, shows that every wind has a tendency to be deflected to the right.

Consequently the winds flowing from high latitudes into lower are deflected, from being Northerly or Southerly winds, into North-easterly or South-easterly respectively. In the northern hemisphere, therefore, we have a North-east Trade, and in the southern a South-east Trade, as the respective under currents.

Currents flowing in the opposite direction, from the equator into higher latitudes, receive a tendency to move from W. to E. and the upper counter-currents flowing from the equator towards the poles, of which I have spoken, the 'Anti-trades,' as Sir J. Herschel so well named them, appear as S.W. or N.W. winds, according to the hemisphere.

In the northern hemisphere these winds blow with great persistency only over the oceans, as the influence of the great continents, with their extensive changes of temperature, from summer to winter, disturb the regularity of the motion of the lower currents. In the southern hemisphere, however, where there is, comparatively speaking, no land to interfere with them, they attain such a force and development as to give the name of 'the roaring forties,' to the belt of latitude in which they are chiefly felt. This exceptional strength of the North-west wind in southern latitudes is unquestionably connected with the permanently low barometrical pressure which is found near the south pole and the cause of which we shall endeavour to explain (p. 251).

The circulation of the atmosphere which I have described undergoes considerable modifications in the course of the seasons. The Trade-wind zones, and the belt of Calms between them, shift their position with that of the sun, or rather with the change of temperature of the earth's surface. In March, when the sea-surface-temperature is lowest, the North-east Trade does not extend as far north by nine degrees in the Atlantic as in September, when the sea-temperature is at the highest. The following table gives the limits of the two Trades and the intervening belt of Calms at these two extreme months:—

	March		September	
	Atlantic	Pacific	Atlantic	Pacific
N.E Trade	26° N. to 3° N.	25° N. to 5° N.	35° N. to 11° N.	30° to 10° N
Calms	3° N. to Equator.	5° N. to 3° N.	11° N. to 3° N.	10° N. to 7° N
S.E. Trade	Equator to 25° S.	3° N. to 28° S.	3° N. to 25° S.	7° N. to 20° S

In March there is a break in the South-east Trade in the Pacific in the neighbourhood of the Friendly Isles.

We have, therefore, in each hemisphere four regions or belts of winds.

I. The equatorial calm belt with variable winds close to the equator.

II. The belt which is constantly in the Trade wind throughout the year.

III. The belt which is in the Trade-wind in summer, and outside it in winter.

IV. The belt of variable winds, principally Westerly, which lies on the polar side of the Trade-wind region.

DISTRIBUTION OF ATMOSPHERIC PRESSURE. 247

Were the earth entirely covered with water, this arrangement would prevail all over the globe, but the influence of the continents on the wind is strongly marked. The land, when its temperature is high, causes an influx of air from the adjacent seas, and when its temperature is low the reverse action takes place, and the air flows out from the land to the sea.

The district where this dependence of air-motion on temperature is exhibited on the most extensive scale is in Southern Asia. It is there that the greatest annual change in the position of the most highly heated area takes place. This is transferred from Persia and the plains of the Punjaub to North Australia, from 30° N. to 20° S., and here we find the true Monsoons blowing for half the year in one way and for the other half in the opposite direction.

In our summer, the S.E. Trade is drawn across the line to the heated plains of Northern India, and the other regions in Southern Asia, lying in about the same latitude. The air current thus drawn across becomes subjected to the same influences as the Anti-trades, already described, and is diverted to the westward like those winds. It blows between May and October over the Northern Indian Ocean and along the coast of China, even as far up as the Amur, as the S.W. Monsoon, although its direction is not always true South-west. In winter, the N.E. Trade holds its course unchecked over the same district.

At the other side of the equator, in the Dutch East Indies, the S.E. Trade prevails at the season when the S.W. Monsoon is felt in India, and it is in its turn, in our winter, replaced by the N.W. Monsoon, formed by the N.E. Trade drawn across the equator.

Thus over the whole of this region the winds are exactly reversed every six months. The peculiarities of these winds first attracted the notice of Europeans when Alexander's troops reached India. Marco Polo first heard of them at Mangi (Burmah), whence the boats sailed yearly to Zipangri (Ceylon) with a fair wind, returning in six months' time with a wind as fair in the opposite direction. The term 'monsoon' is a corruption of an Arabic word 'mausim' (pronounced *mausum*), signifying season.

These are, in very general terms, the systems of constant, or periodical, winds prevailing at the earth's surface over about thirty degrees on each side of the equator, and therefore over half the extent of the globe; for in each hemisphere the area comprised between the equator and lat. 30° is 2,308,000 square miles, while that between lat. 30° and the pole is 2,323,000 square miles.

I must now say a few words about the general laws which regulate the variable winds over the temperate and frigid zones, and the discovery of which may fairly be attributed to the isobaric charts of Buchan.

These winds are governed generally by the distribution of atmospherical pressure, and, locally, by the configuration of the ground. The former of these agencies brings us to the consideration of Plates VI. and VII., which represent the distribution of pressure in January and July respectively. To understand these charts, we must say a few words as to the general reasons why the barometer rises or falls, stands high or low.

If we leave alone the question of storms and the violent changes of pressure which accompany them, we may say with Prof. Mohcr[1] that the barometer stands high—

[1] *Grundzüge der Meteorologie*, p. 206.

(1) When the air is very cold, for then the lower strata are denser and more contracted than when it is warm. The contraction causes the upper layers to sink down, bringing a greater number of air particles, that is, a greater mass of air into the column of which I have spoken before (p. 242), so that the pressure at its base is greater.

(2) When the air is dry, for then it is denser than when it is moist.

(3) When in any way an upper current sets in towards a given area, for this compresses the strata underneath.

Conversely, the barometer stands low—

(1) When the lower strata are heated, causing the surfaces of equal pressure to rise, and the upper layers to slide off as already described, for by this means the mass of air pressing on each unit of area below is reduced.

(2) When the air is damp, for as the density of aqueous vapour, at the temperature of 60° and pressure of 30 ins., is $= 0\cdot622$, air being $= 1$, the mixture is lighter the more vapour it contains, and consequently damp air does not press so heavily as dry on the unit of area below.

(3) When the air from any cause has an upward movement, for this of course acts in the same manner as (1).

We shall not find it difficult to account on these principles for the position of the areas of permanently high and permanently low barometrical pressure.

A favourable condition for the production of exceptionally high barometrical readings over a country is the presence of deep snow, as Woeikof has shown; for if snow is absent during a period of frost—

1. There is no non-conducting stratum interposed between the soil, which is comparatively warm, and the cold air above, so that the thermal exchange proceeds unchecked.

2. The moisture of the soil tends to promote thermic equilibrium between the earth and the atmosphere.

3. The ground, being unprotected, is rent by deep cracks, into which the air penetrates and, becoming warmed by contact with the soil, rises to mingle with and warm the atmosphere above.

Accordingly, in the absence of snow the cold cannot attain such a severity as when snow is lying. If, then, the cold is not so great the lower strata will not be so contracted and the barometer will not rise so high.

The regions of highest barometrical readings are, therefore, situated over the continents and in high latitudes, and they occur in localities characterised by abnormally low temperatures.

Conversely, if the air over any region is very warm and damp, there will be generated a tendency to an upward movement which will cause the barometer to fall, and accordingly we find an area of low pressure in winter in high latitudes in the northern hemisphere over both the Atlantic and Pacific Oceans, where we have already seen from Plate IV. that the temperature is abnormally high.

There are, however, two great facts, as to the distribution of pressure and atmospheric circulation, which will appear on Plates VI. and VII., and of which a satisfactory interpretation has not yet been given. I speak of the exceptionally low barometrical readings near the south pole and of the strong Westerly winds of the southern hemisphere.

The following theory, however, proposed by Andries,[1] being based on the general principles of atmospherical circulation already stated, appears to offer some explanation of these phenomena.

We have already seen in Chapter XII. that though, as Dove showed, the southern is on the whole the colder hemisphere, in all probability the southern hemisphere beyond the parallel of 50° is at a higher temperature than the northern in corresponding latitudes, owing to the comparative insignificance of the land area to be found in the southern. Beyond 40° S. there is practically no land, with the exception of the Antarctic continent, the precise area of which is quite unknown. There are, however, reasons for the belief that within the parallel of 60° there is really much less land at the south than at the north pole. In the latter case the circle of latitude passes almost entirely over dry land, while in the former its course is all but exclusively over water.

The result of these conditions of greater heat in the south is to cause the atmosphere to be more rarefied in that hemisphere than in the other, or to raise the surfaces of equal pressure, so that the differences of pressure in the upper strata between the equator and the higher latitudes will be less, and consequently the force tending to produce an overflow towards the south from the equatorial ascending mass will not be as great as it is towards the north.

Again, from the comparative absence of land in the

[1] *Zeitschrift der Oesterreichischen Gesellschaft für Meteorologie*, vol. xv., 1880, p. 53.

southern hemisphere, there will not be so favourable an opportunity for the production of districts of high pressure and consequent generation of indraught currents from above, as is afforded by the extensive land areas of Northern Asia and British North America.

The mean pressure of 29·4 inches, which in our hemisphere is only found in the neighbourhood of Iceland in January (Plate VII.), extends in corresponding latitudes in a belt all round the south pole. The reduction of pressure becomes even greater farther south, but, as to this, the deficiency of observations prevents our being able to speak positively. So much, however, comes out of the records of Sir J. Ross's voyage, in which he penetrated to 80° S., on about the meridian of New Zealand, that in the months of January and February the mean barometer was lowest, 28·898 inches, between lat. 70° and 75° S., and began to rise as the expedition proceeded farther south, the winds experienced affording independent evidence that the conditions of pressure were changing.[1]

I have said that there is a general deficiency of land in the southern hemisphere, but there is a considerable land area near the south pole, and the Antarctic continent must act in the same manner as Arctic polar lands, and constitute a centre of low temperature—a pole of cold, and, consequently, must act as a centre of indraught for the upper currents from the warmer water surfaces about it. This action tends still further to reduce the pressure over the region at and about the Antarctic Circle.

[1] *Contributions to our Knowledge of the Meteorology of the Antarctic Regions.* Published by authority of the Meteorological Committee, London, 1873, p. 9.

Recognising, however, the existence of this great defect of pressure, it is not unreasonable, on the principles of the relation of wind to barometrical pressure, which I am just about to explain, to account for the exceptional force and frequency of the Westerly winds, which form the most prominent feature of the climate of the south temperate zone. For instance, we hear that, out of the sixty-eight days during which Sir J. Ross lay at Kerguelen, a gale from North-west was reported on forty-five. We have our 'brave west winds' at this side of the equator, but they are far from blowing with such persistent force as that experienced at Kerguelen.

I have just said that the winds met with in very high latitudes near the south pole showed that the conditions of pressure were changed, for the great predominance of Westerly over other winds had disappeared; and this brings me to the subject of the relations of wind to barometrical pressure.

Wind, of course, arises from differences of pressure, as the air must move from a region where the barometer is high towards one where it is low, in order to restore atmospheric equilibrium. This motion, however, does not take place in a straight line from the region of high to that of low pressure, because the same force which we have mentioned when speaking of the Trades and Anti-trades comes into play. The air is constantly drawn away to the right in the northern, to the left in the southern hemisphere, by a force generated by the earth's rotation on its axis. I have already explained how in the northern hemisphere Northerly winds tend to become North-east and Southerly to become Southwest. In fact, as Dove says, most winds are storytellers, for they do not really come from the point from which they are blowing.

A little reflection will show that, if we have currents of air setting in from all sides towards one spot, and all caused to deviate in one way, the result is that a gyratory movement must be generated round that spot. This gyratory movement in the northern hemisphere must leave the point of attraction, the spot of lowest pressure, on the left-hand side. Conversely, if the air flows out in all directions from a point where the pressure is high, a gyratory movement must be developed which will be in the opposite direction to that just described, and will leave the point of highest pressure on its right-hand side.

From these considerations we gather that round an area of low pressure in the northern hemisphere the wind will circulate, having the lowest pressure on its left, or in a direction against the hands of a watch. Round an area of high pressure in the same hemisphere it will circulate in the opposite direction, or with watch-hands.

In the southern hemisphere these conditions will be exactly reversed; the wind will move round an area of low barometer readings with watch-hands, and round an area of high readings against watch-hands.

These principles of the relation of wind to pressure hold good absolutely, without exception, and are to be found in the pages of older writers.[1] They were definitely stated by Coffin and Ferrel in America more than twenty-five years ago. In Europe, Professor Buys Ballot, of Utrecht, first drew attention to their

[1] *E.g.* Prof. Adolf Erman in 1853 in Poggendorff's Annalen vol. lxxxviii. p. 260.

DISTRIBUTION OF ATMOSPHERIC PRESSURE. 255

importance, and the law expressing them has received the name of Buys Ballot's Law. It is generally stated as follows :—

'In the northern hemisphere, stand with your back to the wind and the barometer will be lower on your left hand than on your right.

'In the southern hemisphere, stand with your back to the wind and the barometer will be lower on your right hand than on your left.'

This law shows us that, knowing the distribution of pressure, we can tell the direction of the wind, and *vice versâ*. We shall have to refer to it repeatedly when we come to treat of weather and storms; at the present time we content ourselves with pointing out that the relations represented on Plates VI. and VII. show that it holds equally good for mean winds and mean barometrical pressures as for the comparatively ephemeral conditions of weather and storms for which Professor Buys Ballot stated it. In the northern hemisphere, wherever we find an area of low readings, the wind moves round it against watch-hands, and wherever we find an area of high readings the wind moves round it with watch-hands. In the southern hemisphere the reverse holds true in both cases.

Let us now consider the conditions of pressure on Plates VI. and VII., for January and July respectively. It must be understood that these maps do not represent the actual average readings which are met with at the respective stations over which the lines pass, as the readings are reduced not only to 32°, but to sea-level, in order to make them comparable with each other.

Moreover, in the case of many of the lines the information requisite to draw them with perfect confidence is wanting, either from deficiency of barometrical readings in uninhabited regions or uncivilised countries, or from ignorance of the data requisite for the correct reduction of the readings to sea-level; for, as will be seen from Chapter V., p. 86, this reduction depends on several distinct conditions, some of which it may be impossible to determine with certainty.

Both maps exhibit, speaking very generally, five belts of pressure. Firstly, a band of low readings lies near the equator; on either side of this we have zones of high pressure, where the Anti-trade descends to the earth's surface; and, lastly, outside these we have low pressures in the polar regions.

Examining, however, Plate VI., for January, more particularly, we see that the actually highest readings on this side of the equator are in Eastern Siberia, where the pressure is as high as 30·4 inches.[1] A second region of high pressures is found over the Western States of North America, about the fortieth parallel; and a third, with similar readings, occurs over the Eastern Atlantic, between the parallels of 25° and 35°.

In the southern hemisphere the most remarkable area of high readings lies over the Pacific off the coast of South America in latitudes corresponding to those just mentioned; another, less strongly marked, is found in a similar position in the South Atlantic, and extends round the South of Africa to the Southern Indian Ocean.

[1] Owing to the small scale of the map the isobars for *even* tenths of an inch are the only ones given. The pressure in Eastern Siberia is really above 30·5 inches in January.

The areas of low pressure outside the torrid zone occur in various localities. That which is by far the most conspicuous lies over the North Atlantic between Iceland and Cape Farewell, and in it the readings are an inch lower than those over the Siberian area of high pressure.

The North Pacific, too, has its area of low readings, with the minimum as low as 29·6 inches, stretching across from Alaska to the coast of Kamtschatka.

In the southern hemisphere, where we have already seen that but little contrast of temperature exists at the summer season, there is equally little disturbance of the conditions of pressure, and the chief feature of the map in low latitudes is the existence of an area of slightly defective pressure over each of the three continents which extend into southern latitudes, the principal being over Africa, where the mean readings are below 29·8 inches. The great Antarctic depression then appears, and the isobaric lines are drawn nearly parallel to the circles of latitude, between 40° and 60° S.

If we compare Plate VI. with Plate IV., showing the isabnormals of temperature in the same month, the strong similarity between the general appearance of the curves at once attracts notice. The areas of high pressure are found in positions corresponding to those of negative anomaly, of excessive cold, both in Asia and in North America. In the same way the areas of low pressure coincide approximately with those of positive anomaly, of abnormal heat, in the North Atlantic and North Pacific.

Speaking more precisely, the regions of excessive heat are situated on the eastern side of the areas of low pressure—a relation which we shall find to exist in the case of the ordinary barometrical depressions, which will be described in Chapter XVIII.; and the regions of excessive cold lie also on the eastern side of the area of high pressure, thus further carrying out the analogy. These relations have been pointed out lately by Prof. Wild and by M. Léon Teisserenc de Bort.

If now we look to the general course of the winds as shown on the Plate, we find all round the northern hemisphere a prevalent North-east wind between the high pressure areas over the tropic and the equatorial low pressure.

In the southern hemisphere the conditions are similar. The South-east Trade covers a broad belt from the regions of high pressure up to and beyond the equator, the two principal exceptions to that statement being found, firstly, in the Indian Ocean, where the North-west Monsoon prevails in the Dutch East Indies in January; and, secondly, in the Western Pacific, where the South-east Trade is interrupted by a calm region near the Fiji and Navigator's Islands.

On the polar side of each of the tropical areas of high pressure the West winds appear, precisely according to Buys Ballot's Law. In fact, round the areas of high and low pressure respectively, in high latitudes, the winds circulate in strict conformity with the law, blowing from South-west and West along the western coasts of the continents, and from North-west, North, and North-east along the eastern.

In the South Atlantic and South Pacific, over the open sea, there is a decided circulation of the winds round the respective areas of high pressure, and beyond these we find the strong Westerly and North-westerly winds of which so much has been said already.

Plate VII., for July, is not characterised by features of the same prominence as those just described, but it will be remembered that the thermal conditions of the globe are far more equable at this season than in the northern winter. The isabnormal curves, accordingly, on Plate V., are but few in number, as the contrast in temperature between land and water does not attain so great proportions as on Plate IV.

The chief feature of interest in the plate for July is the great change of pressure over Asia, resulting in the appearance of an area of rarefaction over the centre of the continent, coincident, to some extent, with the closed curve of 10° positive anomaly shown on Plate V.

The difference in pressure over Central Asia between winter and summer is clearly shown by the annual range curve for the barometer at Barnaoul (fig. 19), p. 93.

In the same way the western Territories of North America are marked by an area of low readings in nearly the same position as that of high readings exhibited in Plate VI., but the mean pressure over Utah is at least 0·2 inch greater than that over Western Tartary.

The areas of high pressure over both the northern oceans are very clearly defined, and the readings at their centres are almost the same, viz., about 30·2 ins.

In the southern hemisphere the belt of high pressure, above 30·0 ins., stretches right round the globe, and is marked by five distinct patches of higher readings which lie respectively over the South Atlantic and South Pacific just beyond the Tropic, and over the three continents, which—July being the winter of that hemisphere—become regions over which the air from the comparatively heated surfaces of the adjacent oceans descends.

The southern depression is represented in somewhat higher latitudes, for the isobar of 29·4 ins. lies nearly in latitude 60° off Cape Horn, the position taken by that of 29·2 ins. in January.

Let us now look at the winds. The temperature in the northern hemisphere being higher than in January, the Trades are almost at their highest northern extension, and the South-east Trade crosses the equator both in the Atlantic and Pacific, causing a South-west wind, or Monsoon, over a small area of the sea near the west coasts of Africa and America, while in the Bay of Bengal and Arabian Sea we find, as already stated, the South-west Monsoon up as far as 20° N.

As the areas of high and low pressure over Asia and the North Atlantic respectively have disappeared the strongly-marked circulation of the air induced by them is no longer observable. On the other hand, in the southern hemisphere, July being almost mid-winter, the North-west winds are much more developed in force and extension than is the case in January.

This will suffice to give some idea of the broad features of the relations of wind to pressure over the globe. In the next two chapters I shall endeavour to deal with winds more minutely, and to show how they depend on temperature, and in their turn affect ocean currents and sea-surface temperature.

CHAPTER XIV.

THE PREVAILING WINDS.

In the last chapter a general explanation has been given of the causes of atmospherical circulation, as exemplified by the phenomena of the Trades and Anti-trades. It will now be of interest to enter more into particulars as to the prevailing winds experienced in various parts of the earth, and as to their relations to the principal ocean currents and to sea-surface temperature. These latter relations will form the subject of the next chapter.

Following Dove, I shall divide these winds into three classes—Permanent Winds, Periodical Winds, and the Variable Winds of high latitudes.

The *permanent* winds are the Trades and Anti-trades, which prevail all the year round over a great portion of their respective domains. As we have already seen, the limits of the Trade winds change their position with the season, and, accordingly, at both sides of each Trade-wind belt there are regions over which, for part of the year, the Trade extends, while for the rest of the year they lie outside its area; but over immense tracts in both the N. and S. Atlantic, the N. and S. Pacific, as well as in the South Indian Oceans, in what is called the Heart of the Trade, the wind blows more or less steadily as a North-east or

South-east wind respectively, according to the hemisphere, day after day, throughout the year. On approaching land, wherever that land may be, local influences come into play, affecting the force and direction of the wind, and at times even reversing the latter altogether.

It would be beyond the limits of a work like this to give the monthly changes in the geographical position of the Trade winds; still, it may be well to remark that, during the first nine months of the year, the equatorial limit of the North-east Trade in the Atlantic lies in a higher latitude near the west coast of Africa than it does farther to the westward, until the fortieth meridian is passed, when the limit again recedes from the equator. In the months from October to December, however, the Trade wind is in the lowest latitude on the African coast. The South-east Trade extends farther into north latitude in our summer (July to September), when from 25° to 45° W., its northern limit is about on the parallel of 5° N., and the intervening district of calms is at its narrowest for the year.

As to direction, too, that is modified with the longitude. Along the north-west coast of Africa the Trade is distinctly North-north-east inclining to North, while on the opposite side, along the West India Islands, the direction is Easterly, and even something South of East. These directions exhibit a distinct tendency in the air to circulate round the area of high barometrical pressure already noticed near the tropic of Cancer.

In the South Atlantic the South-east Trade blows with a greater steadiness, and over an area more extensive, than the corresponding wind in northern latitudes. It is decidedly more Southerly on the

African side than farther out at sea, and, even as far south as the parallel of 20°, the Admiralty Wind and Current Charts show that, along the coast, right up into the Gulf of Guinea, the influence of the land diverts the true motion of the air and produces a local Monsoon.

On the American side, however, south of Cape St. Roque, we do not find northing in the wind, which is represented as blowing true East, on to the coast.

In the North Pacific the North-east Trade is felt over an enormous extent of longitude, but its force is not equal to that experienced in the Atlantic Ocean. Its change of direction is strongly marked, being Northerly on the American shore, due East in mid-ocean, while it exhibits a distinct South-easterly direction at certain seasons near the Philippine Islands, where it comes within the influence of the summer Monsoon system of China.

For the South Pacific the Admiralty Charts, already referred to, embody nearly all that is certainly known, and are the main source of information. They indicate during all months, except from July to September inclusive, an interruption in the Trade wind, which stretches in a north-west and south-east direction, and has a breadth of about twelve degrees in the first half of the year, the centre of the interruption lying near the Society Isles, in 20° S. and 150° W. In the three winter months of that hemisphere, July to September, the break disappears, and the Trade wind stretches across the ocean; and in the three spring months, October to December, the interruption between the two parts of the South-east Trade is not nearly so marked as it is from January to June, and it lies more in an east and west direction.

In the South Indian Ocean the South-east Trade prevails throughout the year. On the western coast of Australia it is distinctly more Southerly, and on the African coast more Easterly, than it is midway between these continents. On both sides, however, during the summer months of the southern hemisphere, the direction of the Trade near the coast is changed, under the influence of the Monsoon action.

In all oceans the Anti-trades or Westerly winds, with Northing or Southing in them according to the hemisphere, prevail on the polar sides of their Trade winds, though not with the same permanency as these latter, and extend far up towards the poles. They form a predominant feature of the meteorology of the North Atlantic, but in the Southern Ocean they attain a force and prevalence unknown at this side of the equator. These are the winds which blow in the 'Roaring Forties' of the South Indian Ocean, and whose domain extends rully up to 50° S.

We now come to the *periodical* winds, and of these by far the most important are the great Monsoons of the Indian and China Seas, known popularly as the North-east and South-west Monsoons, but more correctly termed by Blanford in his 'Vade Mecum,' the Winter and Summer Monsoons, for reasons which will at once appear.

Throughout the whole of this region, as far as 30° N. in India, and 20° S. on the coast of Australia, the wind is reversed every six months. During our winter, from October to April, the North-east Trade blows down to the equator, with clear weather. It then crosses the Line, and blows as the North-west

Monsoon up to the limits of latitude already mentioned, bringing with it sultry damp weather, with torrents of rain. In our summer months, April to October, the conditions are reversed ; the South-east Trade extends up to the equator, and, passing to the northward of it, becomes the South-west Monsoon, with rain. In both cases it is the wind blowing from the equator which brings the rain.

Dove's explanation of the South-west Monsoon is, that it is produced by the great rarefaction of the atmosphere over Central Asia in summer, evidence of which is afforded by the contrast in the course of the isobars in January and July respectively (fig. 19, p. 93, and Plates VI. and VII). Later examination of the subject by Blanford and others have shown that this explanation is not sufficient, for the Himalayan range appears to offer an insurmountable obstacle to the northward advance of the South-west Monsoon, and, as a fact, there is no evidence of any wind blowing across the range into Thibet. The regions towards which the air is drawn in each district are situated nearer to the coast than the actual central part of the continent. Blanford says : 'At the former season (the Winter—the North-east Monsoon) the Upper Punjab is, in general, the seat of the highest pressure and the lowest temperature; in the rains (the Summer—the South-west Monsoon) it is that of the lowest pressure and the highest temperature.'[1]

The direction of the wind is not all strictly Southwest, but is locally modified, blowing from South-east up the Ganges Valley towards the heated plains of the Punjab. Along the coast of China the direction is

[1] *Indian Meteorologists' Vade Mecum*, Part II., p. 73.

more Southerly, and farther north it becomes even Easterly, as the region towards which it is attracted lies to the westward. Out at sea, over the Pacific, the North-east Trade prevails throughout the summer months.

Just as the South-west Monsoon is not always a South-west wind, its counterpart in the opposite hemisphere, the North-west Monsoon, changes its direction locally. Near the Mauritius and Madagascar it blows from North and North-north-east, while on the north coast of Australia the direction is West-north-west, and even due West in the narrow passage of Torres Straits.

These winds first became known to Europeans when Alexander's troops reached the shores of the Indian Ocean, and Aristotle described the regular alternation of the wind-currents very accurately, as has already been stated. The Arabs had a very detailed knowledge of these changes, so important to their navigation, for in a work by Sidi Ali on the navigation of the Indian Ocean, published in 1554, the time of commencement of each Monsoon at fifty different places is given.

The periods of change of direction of the current are called the 'Breaking of the Monsoon,' and it is principally at these times that the violent storms of the East Indies occur which, as we shall presently see, are precisely similar in character and effects to the West India hurricanes.

There is also a so-called Monsoon in the Atlantic Ocean along the coast of Africa, from Cape Verde southwards, which, however, does not change with the season. These winds are called by Dove the West

Monsoons of the Line, although the direction is generally from the Southward, and they are not confined to the neighbourhood of the equator, as they blow all along the coast from Cape Verde down to Walvisch Bay, shifting their position somewhat with the time of year. In the Gulf of Guinea, particularly, they are often interrupted by calms, so that the wind does not attain the persistency of the Indian Monsoon.

The coasts of South America, too, on both sides of the continent, exert a decided influence in deflecting the winds towards the land, at certain seasons, but this does not deserve the name of a Monsoon action.

We come, lastly, to the *variable* winds; and I shall first say a few words about those prevailing in low latitudes, before I deal with the winds experienced, for instance, in Western Europe, during all seasons of the year, which in everyone's experience fully deserve the name of 'Variables.'

The Variables, properly so called, are the winds experienced along the region interposed between the two Trades in the Atlantic and Pacific Oceans. In both these localities, near the equator, there is a region of calms, which does not extend right across the oceans, as shown in old wind maps given in Physical Atlases, but is broadest on the eastern side, and thins out towards mid-ocean, or on the western side; the two Trades in the Atlantic apparently meeting and blending together, being attracted by the land of South America, and in the Pacific both currents becoming merged in the Monsoon system of the East Indian Archipelago. In the Indian Ocean, during the northern winter, when the North-west Monsoon prevails to the Southward of the Line, the Admiralty Wind and Current Charts show that calms are extremely prevalent in a broad belt near the equator, and over many degrees of longitude.

On the polar sides of the Trade winds in all oceans, where, as shown in Plates VI. and VII., the regions of high and uniform barometrical readings are found, and where, according to the theory in Chapter XIII., the air overflowing from the torrid zone descends to the sea surface, calms again occur, called by Maury the Calms of Cancer and of Capricorn. In the North Atlantic this region has got the name of the Horse Latitudes, a title which Mr. Laughton derives from the Spanish *El Golfo de las Yeguas*, the Mares' Sea, from its unruly and boisterous nature—as the calms are frequently interrupted by gales of wind from different quarters—in contradistinction to the Trade-wind zone, *El Golfo de las Damas* (p. 244), so called from the pleasant weather to be met with there.

Round each of these regions of high barometric pressure the wind circulates according to the law already explained, with watch-hands in the northern, and against watch-hands in the southern hemisphere; but the actual regions themselves are to be avoided by the seaman, as they may probably cause serious and inconvenient delays, from the baffling and uncertain winds to be met with there.

To the Southward of the Arabian Sea, also, the phenomenon of a calm region during part of the year is found, but it lies close to the equator, not outside the tropic. Blanford says of it ('Vade Mecum,' p. 183):—

'During the height of the South-west Monsoon

there is a tract, lying between the equator and 9° N. latitude, and extending nearly from Ceylon to Socotra, in which the winds are light and the sea smooth. This is known to navigators as the "soft place in the Monsoon," and is taken advantage of, more especially by steamers proceeding westward towards the entrance of the Red Sea.'

Before I proceed to deal with the winds of high latitudes, I cannot better summarise the broad features of atmospherical circulation, as exemplified by the Trades and Anti-trades, and their accompanying Calms and Variables, than by quoting Mr. Laughton's words,[1] with his permission:—

'In each of the large oceanic basins—in the North and South Atlantic, in the North and South Pacific, and in the South Indian Ocean—there is between the tropics a constant wind blowing, in the northern hemisphere, for the most part from between North and East; and in the southern hemisphere, from between the South and East, which in all the oceans is called the *Trade Wind*. Contrary to what might have been expected, contrary to what, indeed, is very commonly asserted, the Trade does not blow in the Pacific with either the strength or constancy that it has in the Atlantic, or in the South Indian Ocean; and in the southern, or most open, part of the Pacific, it is frequently interrupted by Westerly winds, which to the west of the meridian of 165° W., and between the equator and the parallel of 18° or 20° S., prevail throughout the whole of the summer, and occasionally through the greater part of the year.

'On the eastern side of each ocean the Trade wind

[1] *Physical Geography*, 2nd ed., p. 97.

extends farther from the Line, and blows more directly towards it, more directly from the pole of its own hemisphere, than it does in any other part of the basin. On the north-west coast of Africa and the coast of Portugal, as well as on the coast of California, the Trade reaches far to the northward, the extension of it being often met with as far north as 40° or 45° N., and blowing in that higher latitude from N.N.E., N., N.N.W., or even from N.W.; whilst on the west coast of Southern Africa, Australia, or South America, it blows most commonly from between S.S.E. and S.S.W.; and prevails, more especially in summer, as far south as the Cape of Good Hope, Cape Leeuwin, or the southern extremity of Chili.

'Towards the western part of each ocean the Trade wind becomes more Easterly; in the North Atlantic it blows among the West India Islands from East; in the South Atlantic it very commonly reaches the coast of Brazil also from the East; in the Pacific it blows along or parallel to the equator, both to the northward and southward of it, through many degrees of longitude; in the South Indian Ocean it prevails from due East, often for days together; and in all these different localities the direction, while approaching closely to East as a mean, is often from one side or the other of that point—frequently from E.N.E., frequently from E.S.E.

'In North Africa—which may be considered as, to a certain extent, a basin of itself, secondary to the basin of the North Atlantic—the winds appear to follow the same rule. On the eastern side, in Egypt and Nubia, they blow with very great constancy from the North; whereas, in the desert, all our observations, limited indeed in number, show that the prevailing direction of the wind is from the East. This Easterly wind, too, is very frequently felt at sea, blowing hot from the coast of Africa, but the greater force of the Atlantic Trade would seem most commonly to carry it with it to the southward.

'In the extreme east of both the Atlantic and Pacific Oceans, near the equator, there is a wide district where the winds are uncertain and the weather detestable. In both oceans, a branch of the Trade wind from both hemispheres follows the direction of the coast line, or very nearly so, and gradually dies out in a region where squalls, puffs from various quarters, mostly from the Westward, heavy rain, violent thunderstorms, are almost the only interruptions to the stifling, humid calm that is most commonly experienced. And in both oceans this region stretches out to the westward in a triangular form, its base resting on the coast, like a wedge thrust in between the Trades of the northern and southern hemispheres.

'And of the Trades themselves it is to be especially noticed that they blow, more particularly in the Atlantic and South Indian Ocean, with a constancy, regularity, and pleasant genial steadiness, of which no one acquainted only with the uncertain, variable, and frequently boisterous winds of our English latitudes can form an adequate notion. In the Pacific their general character is the same; but they are less strong, less steady, and, as has been already said, are to the south of the Line often very broken for months together. In the South Atlantic the Trade is stronger than in the North. As the result of the comparison of many logs, Captain Maury estimates the average speed of a well-found ship in crossing the Trades of the Atlantic, with the wind abeam, or just abaft the beam, at eight knots in the southern hemisphere, and at six in the northern. In the South Indian Ocean it is commonly stronger than even in the South Atlantic; and everywhere it is stronger in winter than it is in summer, in the South Indian Ocean attaining in winter often to the strength of a fresh gale.

'On the western side of each of the great oceanic basins there is a more or less well-marked alternation of the winds in summer and winter. On the coast of Florida, in the Gulf of Mexico, in Texas, and in others of the Southern States, the wind, which during the winter months blows almost constantly from the Northward, during the summer blows no less constantly from the Southward. An alternation precisely similar takes place every season in the China Seas, an alternation which our rich commerce and our recent wars with the Celestial Empire have rendered more familiar, both to navigators and to the general reader.

'South of the equator, the change is very clearly marked on the coast and over the vast plains of South America. The Trade wind, which during the winter blows to the coast of Brazil, south of the parallel of 10° S., from about E.S.E., in the summer prevails from about E.N.E., or from a point even still more northerly. Inland, up the valley of the Amazon, where the Trade blows nearly due East, the change is to a wind which blows down the river or from the West. Near the east coast of Madagascar the wind in summer blows from the North or North-west, and in the Mozambique Channel from between North and N.E.; whilst in the winter the Trade blows fresh from the

South-east, or, in the narrow channel, from the South-west. There is a similar change on the east coast of Australia, which is shown with sufficient distinctness by a comparison of the numbers of equatorial and polar winds observed in the different seasons, although the unsteady nature of the winds of the western part of the South Pacific renders it less strongly marked there than in other places.

'And wherever this alternation is found, the winter, or polar wind, is for the most part a steady, pleasant breeze, freshening at times into a moderate gale, accompanied by fine clear weather; the summer, or equatorial wind, blows fitfully and in squalls, is unpleasantly sultry and moist, is accompanied by heavy rain, thunder and lightning, and is frequently interrupted by the most violent storms of which we have any knowledge.

'The Basin of India and the Indian Ocean north of the Line is distinguished from the other great basins which we have just been considering, by the fact that this alternation of winds is experienced over its whole area. The Polar or Northerly wind blows during the winter; in places with very marked Easting, as on the coast of Arabia; in places with very marked Westing, as in the plains of India and in the valley of the Ganges; and in some places, as on the Malabar and Coromandel coasts, the Northing itself is the most marked feature of the direction; but everywhere it blows steadily and brings fine weather. The wind of summer, which, with a similar variation of direction in the different parts of the basin, is for the most part a wind from the South, has all the disagreeable qualities of the other summer equatorial winds which have just

T

been mentioned. During its season the weather is bad, heavy rain is almost continuous, squalls are frequent, and violent storms not uncommon ; and, in short, the characteristics of the Monsoons, or alternate winds which prevail over the whole of the Basin of India, do not differ in any noteworthy respect from the winds of those easterly coasts and countries where a similar alternation is experienced every summer and winter.

'In both hemispheres, to the north or south of the parallel of 35° or 40°, a strong Westerly wind blows with great constancy all round the world. In the southern hemisphere, more particularly, it blows with a persistence little less than that of the Trade winds, but with a strength which, although fitful, is very much greater. From a fresh, strong breeze, it rises very frequently into a violent gale, and, as such, blows for days together, the mean direction being very nearly West, from which it seldom varies more than a couple of points on either side. South of the Atlantic, south of the Indian Ocean, south of Australia, in the higher latitudes of the Southern Pacific, and to the southward of Cape Horn, we find it still the same—a Westerly gale, whose strength and constancy combined have enabled Australian clippers to make passages which seem to border on the fabulous. In the northern hemisphere it has not the clear range which it has in the southern; but there too it prevails in the most decided and marked manner. Alike in the Atlantic and Pacific; in North America, west of the Rocky Mountains; in the Eastern States; in European Russia and Germany, and in Northern Asia, we find the same predominance of Westerly winds. If they fail on the eastern side of the mountain ranges which run towards the north and south—of the Rocky Mountains, of the mountain system of the west of Europe, or of the Ural Mountains —they soon recommence, and sweep across the countries on the east of these, with the same fresh and occasionally boisterous strength as across those on the western side.

'In each of the oceans, between the Westerly winds and the Trade—that is to say, about the latitude of 35° or 40° N. or S.—there is a large area where the winds are very unsettled, blowing unsteadily from all points of the compass, or alternating with wearisome calms. Many geographers represent these regions as extending, east and west, from one coast to the other; for this there is no authority whatever; and it is quite certain that the polar winds on all the west coasts, whether of Portugal, Africa, and California in the northern hemisphere, or of Africa, Chili, and Australia in the southern, blow fresh across the belts shown on most anemological maps as the "Calms of Cancer" or "of Capricorn," and form, most distinctly, a junction between the Westerly winds and the Trades; whilst on the western sides of the oceanic basins, equatorial winds are almost constantly met with in those latitudes, and, by the gradual veering of the Trades, seem to be another connecting link, or series of connecting links, between the two great atmospheric currents—the Easterly, or Trade winds, and the Westerly winds of the higher latitudes.'

The winds of the temperate zone and of the higher latitudes seem to be regulated by the distribution of pressure, and this, as we have already seen, seems to depend on the distribution of temperature. The winds circulate round the areas of excessive and defective pressure in opposite directions, and to the respective prevalence of warm or cold winds, and to the ocean currents produced by them, we owe the contrasts of climate which are so marked a feature of countries on the same parallels of latitude, as we shall see when we come to deal with that subject.

When Dove, nearly half a century ago, first calculated thermometric windroses, he endeavoured to explain all the phenomena of climate on the hypothesis of two great wind-currents in each hemisphere, corresponding to the Trades and Anti-trades, which he called *equatorial* and *polar*. He then calculated for various places in Europe the variations of the different meteorological elements for winds from each point of the compass, or, as he called them, the barometric, thermometric, and hygrometric windroses, showing that the barometer was lowest and the temperature was highest when the wind was Southwest, the circumstances being exactly reversed in both particulars when the wind was North-east.

As regards the alleged marked predominance of two winds from nearly opposite quarters, over winds from other points, the following table, showing the relative prevalence of the wind, referred to eight points, in the north and south of the British Isles, in the Orkney Islands and in London, appears to show that the idea has but little foundation:—

	N.	N.E.	E.	S.E.	S.	S.W.	W.	N.W.
Orkney Islands	10	5	9	19	13	11	20	11
Greenwich, 10 years (1861-70)	10	12	7	7	8	33	15	7

The figures for the Orkney Islands are taken from Table IV. of a paper by Dr. Clouston, which appeared in the fifth number of 'Meteorological Papers' published by authority of the Board of Trade, 1861. Those for Greenwich are from Mr. Glaisher's paper 'On the Direction of the Wind at the Royal Observatory, Greenwich, for the Ten Years ending December 1870.'[1] The proportions of the different winds in each month vary very much from year to year, but Table XXVI. in Mr. Glaisher's paper, part of which I subjoin, gives the general summary of the results :—

Average number of hours the wind was :—

	N.	N.E.	E.	S.E.	S.	S.W.	W.	N.W.	Calm
January . .	44·8	70·6	35·2	78·5	77·5	260·7	89·7	33·5	53·5
February . .	55·9	72 5	39·4	38·2	56·3	239·2	106·2	39·5	29·6
March . . .	105·9	137·7	53·3	37·4	40 8	163·8	94·5	67·4	43·2
April . . .	63 4	89·1	89·6	42·0	37 8	185·0	87·5	49·1	76·5
May	74·8	125·1	91·7	51·4	45·6	199·8	69·2	32·8	53·6
June	96·5	80·9	41·3	35·8	33·8	216·9	124·6	62·1	28·1
July	72·9	79·2	43 3	38·5	32·7	249·3	130·6	53 8	43·7
August . . .	65·3	62·5	30·6	34·0	40·6	258·7	130·0	61·5	54·8
September . .	30·3	73·6	42 1	43·0	69·4	270·2	113·4	21·0	51·0
October . . .	61·9	79·1	61·9	70 6	65·9	213·3	95·2	40·0	56·1
November . .	93 1	71·3	23·2	44·0	64·5	237·4	101·6	47·4	37·5
December . .	62·4	77·3	41·8	53·0	76·1	237·1	110·2	49 7	36·4
Sums . . .	827·2	1018·9	599·4	566·4	641·0	2737·4	1252·7	557·8	564·0

Here, then, there is no evidence of special prevalence of North-easterly or Easterly winds.

The study of more copious materials than were at Dove's disposal has shown that the predominance of South-west and North-east winds in Central Europe is a local phenomenon, and that the principle which underlies all wind-motion, *speaking generally*, is that in winter the air flows off the land on to the sea, while in summer it flows off the sea on to the land, and this takes place all round the world. On the east coasts of continents, as of Asia and North America, the pre-

[1] *Quarterly Journal of the Meteorological Society,* vol. i. p. 1.

dominant directions of the opposing currents, wherever these are traceable, are more nearly North-west and South-east than such as Dove assumed them, and it is the South-east wind in winter, the sea wind, which is in such regions the dampest and warmest, not the South-west wind, as with us.

Dr. Hann, whom I have so often quoted, has calculated the prevalent winds and the thermal windroses for a great number of stations in the northern hemisphere, and as regards the Old Continent he has shown that the point of greatest cold in the windrose shifts through 108°, from N. 62° E. (nearly E.N.E.) over the North Sea, to N. 46° W. (N.W.) in Eastern Asia. The warmest point, too, shifts through 61°. In all cases the sea-winds are the rain-bringers, while the land-winds are dry.

In his latest work, his 'Erdkunde,'[1] Dr. Hann gives the following tables, which exhibit the contrast, as to frequency and as to thermal characters, of the winds experienced on the opposite coasts of the continents. The wind being calculated to eight points we get the following results:—

WINTER.

Mean Frequency of the Wind in Percentages.

	N.	N.E.	E.	S.E.	S.	S.W.	W.	N.W.
Western Europe	6	8	9	11	13	25	17	11
Eastern Asia	12	7	6	4	4	9	24	34
Eastern North America	12	11	6	7	9	15	15	25

Thermal Windroses.

	N.	N.E.	E.	S.E.	S.	S.W.	W.	N.W.
Central Europe	−5°·4	−7°·0	−5°·8	+2°·3	+2°·3	+5°·6	+4°·3	+0°·7
Eastern Asia and America	−4°·3	+1°·1	+6°·5	+9°·5	+10°·4	+7°·6	+1°·1	−4°·5

---

[1] *Allgemeine Erdkunde*, von Hann, und Hochstetter, und Pokorny. 3te Auflage. Tempsky, Prague, 1881.

THE PREVAILING WINDS. 279

These show us that in winter in Western Europe the most frequent wind is South-west, while both in Eastern Asia and Eastern North America it is Northwest; but these two latter regions differ in this respect, that though W. comes next to N.W. in both cases, in Asia, S.W. comes far behind, while in Eastern North America it is equally frequent with W.

Looking now at the thermal windroses, which show the amount which each wind raises or lowers the temperature, as compared with the mean value of that element, we find that S.W., the most frequent wind in Western, is the warmest in Central Europe, raising the temperature 5°·6; while N.E., the coldest wind in Central Europe, lowering the temperature 7°·0, is in the West the least frequent but one of all the eight shown in the table; whereas on the eastern coasts of Asia and America the most frequent wind, N.W., lowers the temperature as much as 4°·5, while the South wind, which raises the temperature more than 10°, is almost the rarest of all.

These facts, when taken with those connected with the great heat-carriers of the globe—the ocean currents—and with those of the distribution of rain, both of which I shall describe in the following chapters, are amply sufficient to account for such facts as that the coast of Labrador in January has a temperature fully forty degrees colder than that of the British Isles along the same range of latitude.

The thermal conditions here explained are strictly in accordance with the relations of wind to barometrical pressure, already stated. The wind in the northern hemisphere circulates round the areas of high pressure with watch-hands, and round those of low pressure against watch-hands. We, therefore, find Northerly winds on the eastern, and Southerly winds on the western, sides of the high-pressure areas. As these areas in winter are situated over the land, the eastern coasts of the continents have prevalent Northerly, the western prevalent Southerly, winds. The areas of low pressure in high latitudes over the North Atlantic and North Pacific produce the same result; the wind leaves the lowest pressure on its left-hand side, and so we have Southerly winds on the eastern, and Northerly winds on the western shores of the oceans.

In July the conditions of pressure are reversed. Instead of an area of high pressure over Eastern Siberia, we find one of low pressure over Central Asia, while a corresponding change appears over North America.

Let us now see what the table of wind frequency and the thermal windroses for midsummer tell us:—

SUMMER.

Mean Frequency of the Wind in Percentages.

	N.	N.E.	E.	S.E.	S.	S.W.	W	N.W.
Western Europe	9	8	7	7	10	22	21	17
Eastern Asia	7	9	17	22	16	10	9	10
Eastern North America	8	9	7	10	17	23	12	14

Thermal Windroses.

	N.	N.E.	E.	S.E.	S.	S.W.	W.	N.W.
Central Europe	−0°·2	+1°·6	+3°·1	+4°·0	+3°·1	+0°·4	−1°·8	−1°·8
East of Asia and America	−3°·2	−3°·4	−2°·9	−0°·7	+1°·8	+2°·2	+0°·2	−2°·2

For Western Europe, S.W. is again the point most strongly represented, but W. comes very close to it, and N.W. shows a decidedly increased frequency, compared with the figures for January.

In Eastern Asia N.W. and W., by far the commonest winter winds, have become comparatively rare, and the first place is taken by S.E., the two adjacent points of E. and S. coming not far behind. These facts indicate, in no doubtful way, the result of the Monsoon action on the coast of China, and the extension of the diverted North Pacific Trade-wind to the Asiatic seaboard.

In Eastern North America the conditions of wind frequency are not so entirely reversed as in Eastern Asia, for the commonest wind has only moved to S.W. from N.W., but the N. and N.E. winds are decidedly less frequent than they were in January.

The thermal windroses tell us, what we might have already inferred from Plate II., that the contrasts between the different points, as regards their effect on temperature, are comparatively slight. In Central Europe, S.W., the wind commonest in the West does not affect the thermometer to the extent of half a degree, and the greatest depression, with W. and N.W. winds, is only 1°·8, as compared with 7°·0 in January with N.E. wind, while the greatest elevation of temperature, with one of the least frequent winds in Western Europe, is only 4°.

On the eastern coasts the seasonal change is more striking; we find no longer any wind raising the temperature nine or ten degrees, the warmest, S.W., only raising it 2°·2, while the coldest, which is N.E., instead of N.W., as in January, produces a fall of temperature considerably less than the greatest depression at the opposite season.

When we come to treat of the distribution of rain, we shall see more clearly how the existing differences of climate have their origin mainly in the prevailing winds, which I have attempted to exhibit in Plates VI. and VII.

Such representations are, however, far from satisfactory. Firstly, the size of the map is so small that all local details must be disregarded; but even if the scale were increased to that of the largest general wind charts which have been published, the 'Wind and Current Charts for the Atlantic, Pacific, and Indian Oceans,' we should not attain a perfectly unexceptionable representation of the winds actually felt on the earth's surface. Secondly, the tables which have just been given show that for different regions the proportion of winds from different points of the compass varies, and we should require a windrose, or eight-pointed star, corresponding to every arrow given on the chart.

As with temperature and pressure, so with the wind, the material is defective. We can obtain but little information from the hearts of the continents, especially in the southern hemisphere; and as to the sea the information is only obtainable from ships; moreover, the number of logs which really come into the hands of the compilers of wind charts is insignificant compared with the number of ships at sea, so that the material available for discussion bears no comparison with that which would be furnished by a number of ships anchored in the district whose winds are being studied; while the value of the evidence furnished by ships, which are movable observatories, is materially influenced by the fact of their motion.

In almost all the discussions of wind which have been carried on, up to that for the region near the Cape of Good Hope, which has just been completed under the direction of the Meteorological Council, all the observations contained in a ship's log have been used without discrimination; in other words, no regard has been paid to the question whether the wind was favourable or not to the ship, which, nevertheless, has a material influence on the nature of the evidence derived from her log. Of course, for a given area a seaman has much fewer opportunities of recording a favourable than an unfavourable wind.

To make my meaning clear, I shall take a case of common occurrence in European waters. In the spring time there is frequently a prevalence of Easterly winds at the entrance of the English Channel, between the coast of Ireland and that of Spain. This wind delays every ship bound to the eastward, to a northern European port, while it quickens the run of every ship bound to the westward and southward. Suppose that a well-found ship, A, outward bound, running before this wind, makes five degrees of westing, from the meridian of 5° W. to that of 10° W., in twenty-four hours; for that space of about 300 miles of longitude the captain will enter 'East' as the wind's direction, at four-hourly intervals, six times, and no more, for he has no opportunity of knowing how long the wind has been blowing in that district before he reached the meridian of 5°, or how long it blew there after he had passed the meridian of 10°.

Let us now suppose a sister ship, B, homeward bound, reaching the meridian of 10° on the same day as A reached that of 5°, it may easily take her a week to make the 300 miles of easting, working dead to windward; and during these seven days the captain will enter East as the wind's direction forty-two times.

Now when the logs come to be examined, and all the observations for the month for the area between 5° and

10° W. longitude are extracted, A's log will give only six observations of East wind, while B's log gives forty-two.

If an equal number of ships, homeward and outward bound, sailed over the same area in each month, these discrepancies would counteract each other, but such is not the case. In one month, the majority of vessels may be going in one direction, and in another in the other. In some parts of the ocean, as for instance near the Cape of Good Hope, ships run to the eastward in one latitude and to the westward in another.

In order, in some measure, to reduce the effect of this tendency of observers at sea to enter in a given area more observations of unfavourable than of favourable winds, the Meteorological Council have decided that it will be safest to allow each ship only one observation per day for each single degree square. A fast-sailing ship can easily run through three, or even four, such degrees in the twenty-four hours. Supposing that all her six observations were in one square and from one point, say South, the observation is set down as 1 S. If two were from S., two from S.W., and two from W., the entry would be ⅓ S., ⅓ S.W., and ⅓ W. If she passed through three squares, she would be entitled to one observation for each square.

Such *weighted* observations give a representation of the winds off the Cape of Good Hope different from that to be found in previous wind charts. These latter, for instance, in the month of January give few Easterly winds close to the south coast of Africa, whereas the weighted windrose indicates a strong Easterly wind close inshore.

The fact is that homeward bound ships all run to the westward near the land, because the conditions (wind, weather and current) they find there are favourable for getting to the westward. Now, in summer (January) as already mentioned, Easterly winds *prevail* there, but, being fair winds, they are not so frequently recorded as Westerly winds, which latter detain ships in the locality.

This weighting does not absolutely correct all the error arising from the inequality of the observations, but it tends to reduce its effect.

The preceding considerations are sufficient to show that after all we cannot regard any existing wind charts as affording a perfectly correct representation of the actual facts of air motion over the globe.

In various parts of the world, local circulations, such as land and sea breezes, mountain winds, and local winds are felt, some of which deserve our notice.

Land and Sea Breezes are the best known of the local circulations, because the most frequently met with, especially in hot countries. The general phenomenon is that about noon a breeze sets in from the sea towards the land, and dies away towards sunset. A little before midnight the land breeze commences to blow in the opposite direction, from the land towards the sea, and in its turn is succeeded by a calm about sunrise.

The old explanation of this alternation was that it resembled the circulation generated by a furnace, the air being drawn in from the cooler sea by day, and the reverse action taking place at night. This theory is entirely disproved by the circumstance pointed out two centuries ago by Dampier, that the sea breeze begins in the offing and extends gradually to the coast, while the land breeze comes off from the shore and forces its way out to sea. Dampier's words about the sea breeze are : 'It comes in a fine, small, black curle upon the water, whereas all the sea between it and the shore not reached by it is as smooth and even as glass in comparison; in half an hour's time after it has reached the shore it fans pretty briskly and so increaseth gradually till 12 o'clock ; then it is commonly strongest, and lasts so till 2 or 3, a very brisk gale.'

Mr. Laughton, who has devoted much attention to these winds, speaks of the land breeze 'as coming off in sharp, sometimes dangerous squalls . . . it very decidedly begins from the land, and forces its way out to seaward : lying becalmed in a roadsted, the first intimation of it is, commonly enough, the smell of the land, pleasant or the reverse, as the case may be, and the breeze in noticeable strength follows almost immediately.'

The same author also points out that the phenomenon is not as strongly developed on arid sandy coasts as on shores which are cultivated and covered with vegetation. The south coast of Jamaica is one of the districts where the sea breeze is most strongly marked, and that island is well covered with vegetation.

However, it has been already explained (p. 242) that the heating of the atmospheric column throughout its whole height does not depend merely on the temperature of the surface on which it rests, and that probably the effective heating over the sea is greater than over the land. Mr. Blanford has applied the principles of general atmospheric circulation, to which I have alluded, to interpret the land and sea breeze alternation.[1] His

[1] *Vade Mecum*, Part II., p. 70.

idea is that when the air over the land is expanded by heat and raised, more or less, like a blister, the upper strata slide off towards the cooler sea and produce an increment of pressure at some distance from the land. The air begins to flow from this region of increased pressure towards that where the air is rarefied and pressure is in defect, and so we have a sea breeze setting in from the offing, as Dampier so well describes it, not a wind drawn in by suction and working its way backwards, as would be the case if the particles nearest the heated spot moved first.

At night the action is reversed, the atmosphere over the land is cooled by radiation and contracts; the isobaric surfaces slope towards the land, and the air above slides down from the sea, sinking over the land and pushing its way out as the land breeze.

The sea breeze is generally a damp wind, as it comes from off the water surface, and is therefore charged with moisture. The land wind is essentially dry, and in some places where it prevails is most deleterious in its effects; in Southern India and Ceylon, horses especially suffer seriously if exposed to it.

A somewhat similar reversal of atmospheric motion in the daily period is well known in mountainous countries, except in very high latitudes. In Europe the winds thus arising are known by local names, which are often different in closely contiguous districts: this is especially the case in the neighbourhood of the Italian Lakes. The phenomena are as follows:—About 9 or 10 A.M., a wind, the day wind, begins to blow up the valleys, freshens till the afternoon, and dies down at sunset, when, after an interval, it is replaced by a counter-current, the night wind, blowing down the valley. These are so regular that any interruption in their daily alternation is well known as a sign of coming bad weather, for great atmospherical disturbances interfere with the process which I am about to describe, and which, according to Dr. Hann, constitutes the true explanation of the phenomenon.

The day wind brings up moisture to the upper strata of the atmosphere, and this is condensed, forming caps on the mountain tops and often giving rise to thunderstorms. The night wind, a descending current, carries the moisture down with it, and so the highest peaks are oftenest clear in the early morning.

The reasons of this rhythmical change in air motion are to be sought for in the action of heat. In the daytime the air in the valleys and on the lower slopes of the mountains becomes heated and expanded. The isobaric surfaces over such districts rise, and the air so raised has a tendency to flow towards the mountains and up the upper valleys, as long as the heat action over the lowlands is maintained. At night the temperature in the valleys falls, and the air lying in them contracts, producing a partial vacuum. This causes the air above to descend, so that a downward current is generated, which lasts all through the night.

In speaking of temperature, we have explained how the air is cooled when it rises and warmed when it descends; the change of temperature in the case of *perfectly dry* air for every 183 feet being 1° F. This principle explains the peculiar physical properties of winds crossing mountain chains. When the wind coming in from the sea, and therefore charged with moisture, meets a mountain chain, it is forced to rise; it is cooled by rising, and made to give up much of the vapour it brings with it in copious rains. The result is that the air is rendered dry and cold.

If now the average height of the cols of the chain, above the plain country beyond, be some 4,000 feet, the air in its descent may receive an increment of temperature of over 20° F., and as at the same time its capacity for containing moisture will be increased, it will be felt as a dry hot wind.

This is the explanation of the characteristics of the *Föhn* of Switzerland, and of similar hot winds reported from many other parts of the world. In North Greenland and along Smith Sound, a warm South-east wind has been reported in the middle of winter, whose origin is similar to that of the Föhn, the interior of Greenland being very mountainous. The hot winds of South Africa and of parts of Australia are probably of the same nature, while as to the 'nor'-westers' of New Zealand, at least of the Province of Canterbury, the analogy is absolutely complete.

In Dr. Haughton's 'Six Lectures on Physical Geography' (London, 1881), p. 102, there is a letter from a resident in that colony which contains the following description:—'Strong winds are very prevalent on the plains; that from the "nor'-west" is the strongest and most furious that blows. At the foot of the hills and on the plains it is a very dry, and often a hot, wind unaccompanied with rain; the sky is a peculiar deep dull blue, and any clouds there may be seem not to move. On the tops of the range there rest heavy black clouds which, notwithstanding the furious wind, remain fixed. In the upper valleys very heavy showers accompany the "nor'-wester," the snow melts, and the rivers which rise in the glaciers and upper valleys are very suddenly

U

freshed, rising from 10 to 20 feet in a night. This wind frequently dies away at night and begins again before midday.'

These winds are simply the 'brave West winds' of the southern hemisphere, which bring abundance of moisture from the sea and deposit it in rain and snow on the steep windward slopes of the mountains, descending their lee sides as dry winds and being heated in their descent by the compression. After a period of this parching wind, the relief comes in what is called, in colonial phraseology, a 'southerly buster.' Another colonist, Lady Barker, thus describes the change ('Station Amusements in New Zealand,' p. 5) :—'At last, when our skin felt like tightly drawn parchment, and our ears and eyes had long been filled with powdered earth, the wind dropped at sunset as suddenly as it had risen five days before. We ventured out to breathe the dust-laden atmosphere, and to look if the swollen creeks (swollen because snow fed) had done or threatened to do any mischief, and saw on the south-west horizon great fleecy masses of cloud driving rapidly up before a chill icy breeze. Hurrah, here comes a "sou'-wester"! the parched-up earth, the shrivelled leaves, the dusty grass, all needed the blessed damp air. In an hour it was upon us . . . . The deluge of cold rain came steadily down and we went to sleep to the welcome sound of its refreshing patter.

'All that I have been describing was the weather of the past week. Disagreeable as it might have been, it was needed in both its hot and cold, dry and wet extremes, to make a true New Zealand day. The furious "nor'-wester" had blown every fleck of cloud below the horizon, and dried the air. The "southerly-buster," on the other hand, had cooled and refreshed everything in the most delicious way, and a perfect day had come at last.'

In the same way as the North-west Anti-trade in New Zealand becomes a hot dry wind when it crosses the mountains, the Westerly wind of both Pacific Oceans, when it crosses the mountain ranges which line their eastern shores, comes down as an exceedingly dry wind both in Peru and in the Western States of the Union. In these cases, however, the table land at the interior is at too high a level for the air to gain much in temperature by its descent.

In South America there is a district in about 16° S., called the Punos, which lies near the Pacific coast and is a high table land, at an altitude of 13,000 feet, situated between two ranges of the Andes. It is well known as containing Lake Titicaca, and is about 500 miles in length by 100 in breadth. The mountains on either side are very high; Illimani being situated on the south-east side, and there being several other peaks not much lower than it. The South-east Trade blowing over Brazil deposits any moisture it contains in rain and snow on the eastern flanks of these mountains, and the air, when it descends the western slopes, is so cold and dry that Prescott says the ancient Peruvians preserved their dead by simply exposing the bodies to its action.

To return to the Föhn of Switzerland, the wind which gives rise to it blows on the Italian side of the Alps as a warm Southerly wind, charged with moisture, owing to its contact with the surface of the Mediterranean.

On the African coast, and even in Malta, the South-east wind, called 'Scirocco,' has much of the character of a hot wind, as its temperature is high and it is comparatively dry. It comes off the high land of Africa and descends to the sea level, its essential feature, as regards the sensations of those exposed to it, being the feeling of lassitude which it induces.

A period of this wind is often succeeded by a spell of cold North-west wind, a change which exactly corresponds to that just described as occurring in New Zealand.

The wind contrasted with the Scirocco, along the Adriatic, is known as the 'Tramontana,' a name which explains itself. It is, of course, generally speaking, Northerly. In certain localities it is exceptionally strong and receives special names. The 'Bora' of Trieste and Dalmatia is known as a furious Northerly wind, at the former locality sweeping down off the high plateau of Carinthia. Over Provence, the wind from the same quarter is known as the 'Mistral,' and in the old rhyme

> Le Parlement, le Mistral et la Durance
> Sont les trois fléaux de la Provence,

it is declared to be one of the plagues of the country.

When we come to treat of Weather, we shall see that both of these winds, the Bora and the Mistral, are distinctly related to our own North-west winds, which form the concluding phase of our ordinary cyclonic disturbances, but obtain their exceptional local properties in the South of Europe from the local character of the surface over which they blow.

It would, however, be endless to give all the local names of Mediterranean winds, for these vary in different places. In Italy particularly, all the winds, to at least eight points, have special appellations, as given in the note.[1] On the African coast, the 'Levanter' is an East wind, very prevalent in summer and evidently related to the North-east Trade.

The 'Harmattan,' on the west coast of Africa, is a hot East wind. Coming off the desert, it brings with it clouds of reddish dust which cover the sails and decks of ships far out in the Atlantic. In the winter, when the Trade-wind zone is in its southernmost position, this wind is felt as far down as Sierra Leone and Cape Palmas.

In Spain, the wind coming from the African desert is South-easterly, and is known as the 'Solano'; it is charged with dust, and has all the noxious qualities of the Harmattan and Scirocco, winds which come off heated land. In Egypt, the hot wind from the desert is known as the Khamsin, or 'fifty,' from the idea that it blows for that number of days.

The 'Simoom,' which buries whole caravans in sand, is of the nature of a whirlwind, and will be noticed under the head of 'Storms.'

The deserts and steppes of Central Asia have, in winter, their own terrors for travellers in the 'Buran,' a North-east wind which blows as a gale, is excessively

---

[1] The following are the usual Italian names for the winds:—
N. Tramontana or Settentrione.
N.E. Greco.
E. Levante.
S.E. Scirocco.
S. Mezzogiorno.
S.W. Libeccio.
W. Ponente.
N.W. Maestro.

cold, and brings clouds of drifting snow. This is, however, reported to be far outdone by the 'Purga' of the Lower Yenissei, near Turuschansk, during the prevalence of which the atmosphere is absolutely filled with snow-dust flying before a furious gale ; and the unlucky traveller caught in such a storm cannot advance till it has blown itself out.

Of other winds which have attained a pretty general reputation, usually owing to the dangers to shipping which their prevalence entails, I need only mention the 'Nortes' (Northers) of the Gulf of Mexico, which often cause wrecks in the Bay of Campeachy, on a lee shore. They are related to cyclonic disturbances of the atmosphere over the south-western states of the Union.

On the Brazilian coast, the 'Pamperos' are winds corresponding to the Nortes, and are South-westerly, as they are in the southern hemisphere. They get their name from blowing over the Pampas, and, properly speaking, every South-west wind in those regions is a Pampero, but European usage has generally only applied the term to squally winds, some of which in fact become cyclonic disturbances and whirlwinds on a small scale.

CHAPTER XV.

OCEAN CURRENTS AND SEA TEMPERATURE.

NEXT to the winds in importance to climatology come the Ocean Currents, which are, therefore, most properly considered in connection with the winds. Of late years the origin of ocean currents has been discussed with great fulness by Carpenter and Croll in this country and by Colding in Denmark; but for the purposes of the present work it will suffice to consider them as in the first instance produced mainly by the winds, and, following Major Rennell, the first investigator of their phenomena, to divide them into two great classes —*Drift Currents*, produced directly by the wind, and moving more or less parallel to the wind in each district; and *Stream Currents*, which are produced when drift currents are deflected by meeting with an obstacle, such as a line of coast, and flow on in the new direction until the impetus they have received is expended.

Plate VIII., reduced from the General Current Chart published in the 'Wind and Current Charts' issued by the Hydrographic Office in 1872, will give the reader an idea of the general oceanic circulation, and its broad feature will be seen to be that in all the three oceans, the Atlantic, Pacific, and Indian, there is in the torrid zone a general set of the waters from the east to the west; in fact, there is an equatorial westerly current.

Here, at the outset, the reader must be warned that currents are named according to the direction *towards which they flow*, while winds are named according to the direction *from which they come*.

An Easterly wind produces a westerly drift current, and so the Trades, being Easterly winds, produce a general movement of the surface-water from east to west along the equator.

Commencing with the Pacific, this equatorial current flows till it meets the East Indian Archipelago to the east of New Guinea; it there divides and is deflected, the northern portion, known as the Kuro Siwo, or the Great Black Stream of the Japanese seas, sweeping up into high latitudes outside the chain of islands formed by the Philippines, the Loo Choo group, and Japan, while the southern branch flows on till it strikes the coast of Australia, and there turns eastward again.

In the Indian Ocean the water is embayed on the northern side, so that there is no exit for the stream in that direction. The currents, therefore, in the Arabian Gulf and the Bay of Bengal are purely drift currents, depending on the Monsoons and changing with them.

The main equatorial current is principally developed to the south of the Line, and, when it meets the African coast, turns southward. Part flows outside Madagascar, and there is gradually converted into an easterly current—a change similar to that occurring on the east coast of Australia; but the main portion of the current sweeps down the Mozambique Channel and along the coast, as a strong warm stream, flowing out beyond the south point of Africa.

The Atlantic equatorial current may be said to divide on Cape St. Roque. One portion turns southward along the coast of Brazil and flows on as far as the latitude of the River Plate, where it turns eastward across the South Atlantic. The other, and larger, portion passes northwards, and, combining with the westerly current of the North-east Trade, enters the Caribbean Sea, where it, in its turn, is embayed and turned back on itself. It finally flows forth from the Gulf of Mexico as a rapid stream—the Gulf Stream—which skirts the eastern coast of the United States throughout nearly its whole extent, and then sweeps across the North Atlantic to the shores of Northern Europe.

This rough summary shows us that we meet with westerly currents in low latitudes all round the world, and in every case, as we shall subsequently see, there is a counter-current flowing eastward close along the equator, between the drift currents of the two Trades.

In the middle latitudes the currents are reversed, and are generally easterly. In fact, in each of the five great oceans—the North and South Atlantic, the North and South Pacific, and the South Indian Ocean—great bodies of water are continually circling round in eddies, the equatorial current taking the water westwards, and the return currents, like the Gulf Stream, bringing it eastwards again.

The necessity for the existence of such systems of compensating currents is at once intelligible on the principle put prominently forward by Dr. Haughton in his lectures on 'Physical Geography,' p. 141. It is that ' used by mathematicians in hydrodynamics under the name of the *Equation of Continuity*. This may be expressed very simply by the following self-evident theorem:—*Given a permanent system of constant continuous circulation in any ocean; if any vertical plane be supposed drawn across the ocean, equal quantities of water must cross that plane from right to left, and from left to right, in a given time;* for, if possible, let it not be so, then there will arise a difference of sea level at the two sides of the supposed plane, and the hypothesis of constant continuous circulation will become impossible.'

The equatorial currents are essentially warm water currents, but in high latitudes cold water currents appear. The best known of these is the American Arctic current, which flows out of Baffin's Bay, hugging the coast, as the action of the earth's rotation, which we have already described at p. 245, causes the stream to press against its right bank, which is here the continent of America. The influence of this cold stream is felt as far south as Cape Cod, in lat. 42°.

The North Pacific has no polar current of the same importance as that of the Atlantic, for Behring's Strait, the only outlet of the Arctic Ocean in that region, is too shallow to allow much water to pass through it.

It is, however, in the southern hemisphere, with its immense expanse of water, that the polar currents form a prominent feature of the physical geography. It seems, as Sir F. Evans remarks ('Brit. Assoc. Report,' 1876, p. 175), as if all the surface water between the Antarctic Circle and the parallel of 45° S. is drifting northwards and eastwards. In this latitude it joins with the easterly streams produced by the rebound of the southern branches of the respective equatorial currents, chills them, and appears on the western coasts of America, Africa, and Australia as cold northerly currents, causing the isotherms on these shores to dip down towards the equator, as shown on Plates I.–III.

The best known of these currents, and that standing first on the list, is the Peruvian, or Humboldt's, current, on the coast of South America, but that in the Atlantic is well marked, and on the west coast of Africa the influence of the cold current is such that the temperature of the sea near Cape Town is sometimes 20° lower than in the corresponding latitude on the eastern side of the continent.

I shall now proceed to treat of these currents more in detail, and shall commence with the Atlantic, as the district which has been most studied.

In the first place, there are the two westerly equatorial currents, with the counter-current between them. Of these two the southern is the more important, as it comes from a greater body of water. It strikes upon Cape St. Roque and divides into two branches, of which the northern passes along the north-east coast of the continent and finally joins the northern equatorial. The combined current enters the Caribbean Sea, circulates round the Gulf of Mexico, and finally pours out between Cuba and the coast of Florida as a mighty stream of hot water, which passes the Straits of Bemini with a width of about thirty miles and a discharge of some forty cubic miles per hour, at a rate of over 6 feet per second, according to Colding. It flows on, expanding in width, until it is found off Newfoundland, at a distance of 1,800 miles from Cape Florida, as a stream 320 miles wide, but with its velocity reduced to one-third of its initial amount, while its discharge—the quantity of water moving eastwards—is doubled. It advances farther, to the meridian of the Azores, in 25° W., where, about the latitude of 47°, it bifurcates, one part turning to the right, as Rennell's current, along the coast of Portugal and southwards towards the Cape Verdes, while the other moves onwards past our own coasts, and makes its influence felt even within the Arctic Circle at Hammerfest, in Norway.

The Gulf Stream is essentially a current of hot water. It is generally intensely blue in colour, and is described as a vast river flowing past the American coast and spreading out across the Atlantic like a great pennant waving in the wind. The contrast of temperature between it and the American Arctic current inside it is so great, and the change in passing from one to the other so sudden, that off Halifax a difference of 20°, or even 30°, has been reported between the temperatures recorded during the same day, and a change of 8° or 10° in as many minutes. In fact, along the northern edge, what Bache called the 'cold wall,' the line of demarcation is perfectly definite, owing to the change of colour of the water, and the simultaneous temperatures taken at the bow and stern of the ship, when crossing that line, have been known to differ by several degrees.

The Gulf Stream flows through the Straits of Bemini with a temperature of upwards of 80°; as it moves on it loses much of this extreme heat, but still its effect on the temperature of the North Atlantic is very considerable. Various estimates of the amount of this effect have been made. Among those most commonly quoted is that of Dr. James Croll, F.R.S., who puts it at one-fifth of the total amount of heat possessed by that ocean; so that, were the Gulf Stream non-existent, the average surface temperature of the whole ocean would be $-3°$ F. The details of his calculation will be found in a note.[1]

However, in a recent paper, read before the British Association in 1881 by the Rev. Dr. Haughton, F.R.S., and printed *in extenso* in the 'Report' (p. 451), a very different estimate of the effect of the Gulf Stream will be found. Using Ferrel's tables of temperature, for all latitudes, in both hemispheres, Dr. Haughton gives the following figures for the influence of the Gulf Stream on the climates affected by it in July and January respectively:—

Lat.	July	January
40°	$-3°·4$	$+14°·1$
50°	$-2°·0$	$+21°·7$
60°	$+0°·5$	$+37°·0$
70°	$+0°·5$	$+40°·8$
80°	$-1°·8$	$+19°·4$

On this principle we see that the Gulf Stream actually depresses the temperature in summer, while it certainly raises it in winter. The total annual effect is about one-half the winter effect. Dr. Haughton goes on to show that the result of the stoppage of the Gulf

---

[1] According to Dr. Croll, the Gulf Stream, issuing from the Gulf of Mexico, and flowing in a broad current into the Atlantic, is estimated to convey to that ocean one-fifth of the total heat which is found in it. If we take the temperature of its surface, on the whole, as 56°, and the temperature of space, with Sir J. Herschel and others, at about $-239°$, the total heat of the Atlantic Ocean would correspond to a temperature of 295° (239 + 56). One-fifth of this value is 59°, and if, therefore, we subtract from the assumed temperature 56°, the amount (59°) of one-fifth of the total heat, the residue is $-3°$. In other words, on this hypothesis, the surface temperature of the North Atlantic, without the Gulf Stream, would be thirty-five degrees below the freezing-point.

Stream in July would be *nil*; in January it would produce a general fall of temperature of about 4°.

On our own west coasts the water moves very slowly —not more than an inch or so per second—but that it does move, and carry foreign bodies with it to the coasts of Europe, partly owing to the influence of the prevailing Westerly winds, is proved by West Indian products, such as hard-shelled fruits, beans, &c., being picked up on the western coasts of the British Isles and of Norway.

Rennell's current, which, as we have said, turns to the right at the dividing point on the meridian of the Azores, flows south-eastward outside the Bay of Biscay, turns westwards off the coast of Portugal, and is continued as a south-west current past the Canaries and the Cape Verdes to join the northern equatorial drift, and so complete the circuit.

This circulation is naturally right-handed, owing to the influence of the earth's rotation, and as the winds circulate round the anti-cyclonic region near the tropic of Cancer, so this water flows round and round a broad area where no current is perceptible, and where all drift wood and foreign bodies, which are sloughed off the main current to the right, collect. This region is known as the Sargasso Sea, from the peculiar Gulf-weed, the *fucus natans*, or *sargassum bacciferum*, which grows without a root, covering leagues and leagues of water surface, and affording a home to millions of fish, crustaceans, and molluscs. This weed is not met with in the main current outside.

The water of the Gulf Stream has been estimated to take about five and a half months to reach the coast of Europe from its start'ng at Cape Florida, and, on this hypothesis, the time required for a particle to make the entire circuit round the Sargasso Sea into the Gulf of Mexico again would be about two years and ten months.

One portion of the southerly current off the Straits of Gibraltar moves close along the African coast, and eventually, near Cape Palmas, joins the equatorial counter-current, and flows along into the Bight of Benin, as the Guinea current—an easterly current, along the coast, about 100 miles wide—while outside the equatorial current flows to the westward.

Sir Edward Sabine describes how, in the old slave trade days on the Gold Coast, Captain Clavering, in the 'Pheasant,' in which vessel Sir Edward was a passenger, on his pendulum expedition in 1822, used to lie out in the equatorial current to the eastward of the slave ports, knowing that the inshore current must bring the slavers to the spot where he was waiting for them.

The southerly current off Cape Verde is still a cold current, but its continuation, the Guinea current, is warm, at least at the surface, while a little underneath the water is cold; and in a paper read before the Meteorological Society ('Quarterly Journal,' vol. iv. p. 32) Commander E. G. Bourke, R.N., states that in the month of October the temperature at the surface was 80°, and at the bottom, at the depth of eight fathoms, it was 65°. His attention was attracted to this remarkable difference by the fact of his finding the water used for washing the decks, drawn from beneath the surface by a pump, to be much colder than the water on the surface itself.

The presence of cold water at no great distance from the surface near the equator has also been proved by various observers, who have found that in mid-ocean, close to the Line, the cold stratum wells up, so that the navigator meets with patches of water giving a temperature much below that which might be expected for the season. The charts of Sea Temperature for the equatorial region of the Atlantic, published by the Meteorological Office, show for August the isotherm of 73° in 15° W. and 1° S., with the curves for 75° on each side of it; and on August 3, 1863, Captain Toynbee, between 1° 49' N. and the equator, and between 15° and 17° W., took seven observations of the surface temperature, all below 71°, their mean being 70°·3. This cold water undoubtedly comes up the west coast of Africa, where the low temperature of the surface inshore is a well-known phenomenon, which will now be explained.

The circulation of the South Atlantic is more simple than that of the north, as the coasts are not indented by great gulfs and inland seas, and the islands are so few and so small as to be quite insignificant. The cold water comes up from the Antarctic Ocean and strikes on the African coast near Cape Town, where, as already stated, it makes the temperature about 20° lower than at the corresponding latitude on the east coast, and fills Table Bay with a profusion of fish, for almost all the fishes extensively used for food inhabit cold water. The current flows close along the coast up to the Line, where it turns westward and crosses over to Cape St. Roque, the cold water having gradually sunk below the surface, and only betraying its presence in such spots as those in which Captain Toynbee met with it.

I have already described the course of the northern branch of this stream, which turns off from Cape St. Roque. The southern branch flows down along the coast of Brazil to the mouth of the river La Plata in about 35° S. Here it meets with a cold current from the Southern Ocean, coming up past Cape Horn, and over the region where these two streams are in close proximity to each other, alternations in the surface temperature of 10°, or even 15°, in a day are not uncommon.

This locality, like all others where currents of different temperatures come into collision, is notorious for its fogs, gales, and heavy sea.

After meeting the cold current the Brazil current turns eastward before the prevailing Westerly winds, and flows across to the Cape of Good Hope, uniting on its way with the great easterly drift of the Southern Ocean, and making the circuit complete.

The central area of the South Atlantic has its Sargasso Sea, if we may call it so, where, however, there is little or no weed. The area is marked by the presence of drift wood and other *ejecta*, which are thrown off from the surface of all water in motion, and collect in the comparatively stagnant and currentless parts of the sea.

The currents of the Pacific are less well known than those of the Atlantic, and, the ocean itself being comparatively shallow, and in parts thickly strewn with islands and coral reefs, the motion of its waters is far less free. There are, however, on each side of the equator westerly currents corresponding to the two Trade winds, with the counter-current between them as in the other oceans.

The northern equatorial drift is diverted to the northward by the islands lying outside New Guinea, and gradually turned into a strong north-easterly stream of warm water, whose magnitude is estimated

x

by Dr. Haughton ('Phys. Geogr.,' p. 137) at nearly three times that of the Gulf Stream, though it never attains the velocity of the latter, for its waters are not pent up in narrow channels like those which are met with along the coast of Florida. This is the Kuro Siwo, or 'Black Stream' of the Japanese seas, already mentioned, whose intensely salt water flows on through the ocean like the Gulf Stream on the eastern coast of North America.

The greater portion of the waters of the Kuro Siwo pass across to the American coast, and produce such an effect on the climate, owing to the warmth which they bring with them, that, as Von Baer tells us, on the southern side of the narrow promontory of Alaska you meet with humming birds, while the northern shores, which are washed by the cold current coming out of Behring's Straits, are visited by walruses.[1] It is to this warm current that Sitka and the coasts of British Columbia owe their immunity from the ice which besets the shores of Asia in corresponding latitudes.

The current becomes a southerly one on the American coast, and gradually moves on until it is caught up into the equatorial drift, and the circuit of the ocean is completed.

Maury asserts that there is a Sargasso Sea in the North Pacific, occupying a central position in this circulation, but Laughton doubts the existence of sufficient proof of the fact, and inasmuch as the Central Pacific is rarely traversed by shipping, direct evidence is not easily obtainable.

The Arctic current from Behring's Straits is not of equal magnitude and importance to its corresponding current, which skirts the coast of Labrador, but still the

[1] Dove, *Klimatologische Beiträge*, vol. i. p. 13.

influence of the cold water produced by the melting of the polar ice is strikingly manifested by the loops into which the isotherms of air temperature over the North Pacific are thrown in July (Plate III.) in comparison with their course in January (Plate II.), when the sea in high latitudes is frozen.

In the South Pacific there is no stream to correspond with the Kuro Siwo, carrying the warm equatorial waters into the temperate zone, and the most widely known feature of that ocean is the cold Humboldt's current along the coast of Chili and Peru.

The equatorial current pursues its way westwards between the Tropic and the Line, encroaching somewhat upon the northern hemisphere, until it meets the Navigators, Friendly, and Fiji group, where a portion of it is turned back. The greater part of the stream, however, makes its way to the coast of New South Wales, where it meets, and is reversed by, the Antarctic current from Bass's Straits; the two combined currents sometimes attaining the extraordinary velocity of 100 miles a day, between Australia and the Northern Island of New Zealand. To the south of New Zealand, too, strong easterly currents are found, but the general easterly set of the waters in the middle latitudes of the South Pacific is not so clearly shown as in other oceans. When, however, the drift has once reached the South American coast, it takes a decided course northwards through forty degrees of latitude, and this mass of cold water (Humboldt's current) brings to the coasts of Chili and Peru their sea fogs and cold weather.

In the Indian Ocean, the motion of the water, north of the equator, is entirely regulated by the winds. During the winter half-year, from October to April, when the North-east Monsoon blows, the currents run to the westward round the shores of the Arabian Gulf. During the remainder of the year, when the South-west Monsoon prevails, the water flows in exactly the opposite direction.

The great westerly drift of this ocean lies south of the equator, and flows on till it impinges on the African coast, where it splits into two streams. One turns back outside Madagascar and, after a short southerly course, combines with the easterly drift of the Southern Ocean and flows across to Cape Leeuwin, where it throws up a small branch towards the equator, with the effect of slightly lowering the temperature of the coast of West Australia. The great bulk of this easterly current, however, passes south of the Australian continent, and eventually finds its way into the Pacific, partly through Bass's Straits, and partly round Van Diemen's Land, where, as already explained, the velocity attained is very great indeed.

The remainder of the equatorial drift turns southwards along the Mozambique Channel, being termed successively the Madagascar and the Natal current, until it clears the coast of Africa, where it receives the name by which it is best known—the Agulhas Current—because it runs along the eastern edge of the Agulhas Bank, which extends to the southward of Africa.

The Agulhas current is essentially a warm water stream, and where it meets the south-easterly current changes of sea-surface temperature exceeding 20° in a day are not uncommon, and the region in which they are experienced is almost proverbial among seamen for the violence of its storms and its tremendous seas.

As already stated, there appears to be a general north-easterly set of the waters out of the Antarctic regions and towards the equator, which has been known to bring up icebergs from high latitudes as far as the parallel of the Cape of Good Hope in the South Atlantic. This set in places assumes the features of a strong current, especially to the southward of New Zealand. The same tendency of water to flow out of the frigid zone is exhibited in the northern regions also, as is evidenced by the three drifts of the 'Resolute,' of the 'Advance,' and of the 'Fox,' in the ice, down Baffin's Bay. North of Spitzbergen too the same direction of motion is traceable, for when Sir E. Parry, in 1827, in his sledge journey, reached the latitude of 82° 45' N., he found that the ice on which he was travelling was moving southwards, at about the rate of his own northward advance, so that he had to give up the idea of seeking to penetrate farther towards the pole against such a current.

This current is the commencement of the American Arctic current, the compensation for the Gulf Stream; and the connection between it and the Gulf Stream is noticeable even to the east of Spitzbergen, where the Austrian discovery ship 'Tegethoff' was carried by it in the year 1874, in a direction towards N.W. That objects can be really borne by this current the whole way from Europe to America is shown by the recent discovery (1881) on the shore at Okkak, near Nain in Labrador, of glass fishing-net floats, of the pattern used only by Norwegian fishermen, and which, therefore, must have been washed away from the coast of Norway and have made the circuit round the north of Spitzbergen and along the Greenland coast to the spot where they were found. One of these floats was exhibited at the meeting of the Meteorological Society on March 15, 1882.

The immediate connection between ocean currents and meteorological phenomena is, however, by means of sea temperature. Whether the ocean currents are, as some allege, caused by differences of temperature, or are not, it is undeniable that the different currents are marked by differences of temperature in the water which forms them, and that the masses of warm or cold water with which the sea-surface is overlaid, by the agency of the respective currents, exert a constant and most important influence on the atmosphere in contact with them. Thus we have repeatedly mentioned the influence of the Gulf Stream on the atmosphere resting on the Atlantic; and as we deal with other parts of the sea we shall meet with similar instances.

It is, of course, the temperature of the surface which directly affects that of the atmosphere, and it is practically of the temperature of the surface only that we can have any extended knowledge.

The surface temperature is observed by means of an ordinary thermometer, but temperatures beneath the surface, and even at great depths, may be ascertained by the use of deep-sea thermometers. These must be self-registering, in order to show the extremes of temperature reached, or the temperature at any given depth, and they must also be constructed in a special way, in order to protect them against the pressure of the sea-water, which at considerable depths would compress the bulbs and make the instruments read too high.

The requisite protection is afforded by enclosing the bulb of the thermometer in an outer jacket or casing, partly filled with a liquid, either mercury or spirit. This envelope relieves the actual thermometer from excessive pressure and enables it to indicate correctly the temperature to which it has been exposed.

Temperatures beneath the surface cannot be easily ascertained when a ship is in motion, and the determination of temperatures at great depths, such as those of two or three thousand fathoms, requires special appliances and a delay of some hours, for each observation. Information so collected must, therefore, be fragmentary.

Owing to the high specific heat and low absorbing and radiating powers of water, as compared with dry land, the diurnal and annual range of water temperature, at least in the open sea, is very slight. Near the coasts and in shallow water, however, considerable annual change of temperature is observed. In high latitudes, also, the contrast between the summer and winter temperatures is much more pronounced than is the case between the tropics.

The highest surface temperatures which have been recorded at sea range about 90°, and are met with in localities where the water is comparatively still and undisturbed by currents or winds. It is probable that the strata exhibiting such an abnormal degree of temperature are very thin, and that at a moderate depth much cooler water would be found. The coldest temperature found is somewhat below 32°. The freezing point of pure sea water is about 27°·7, but when salt water freezes, the water has a tendency to crystallise away from the salt that it contains, so that the ice of floes, when melted, yields water much fresher than the sea in which the ice was floating. The fresher the water is the nearer will its freezing point be to 32°, so that when the sea-surface actually freezes its temperature cannot be much below that point.

There is one feature of the behaviour of sea-water at low temperatures about which a misconception has prevailed for many years, and has been repeated in text-books, but which has only been clearly explained of late.

The fact that fresh water has a point of maximum density at 39°·5 is well known, and the phenomena accompanying the freezing of ponds and lakes in winter are explained by that property. When the temperature of the surface falls to 39°·5, the uppermost layer of the water, being heavier than those below, sinks, its place being taken by a warmer, and therefore lighter, stratum. This circulation proceeds unchecked until the entire mass of water in the pond has reached the temperature of 39°·5. Below that temperature the water expands, and there is no tendency in the upper strata to sink; they accordingly grow colder and colder till they freeze, so that when the surface is frozen, perhaps a foot or more in thickness, the water below is still liquid, and may be at a temperature nearly eight degrees above the freezing point.

Up to the year 1840 it was hardly doubted that similar conditions as to changes of density existed in salt water, and it was thought probable that the temperature of 39°·5 prevailed at the sea-bottom all over the globe, wherever the water was not very shallow; this dense stratum being overlaid by warmer water in low, and by colder in high, latitudes. Calculations were even made as to the latitude in each hemisphere in which this critical temperature of 39°·5 would be met with at the sea-surface.

This idea prevailed very generally, in entire disregard of the fact that on several occasions during the earlier Arctic and other exploring voyages temperatures considerably below 39° had been observed to exist at the sea-bottom.

It is a most remarkable fact that although A. von Humboldt, in a letter to the Earl of Minto, dated October 26, 1839,[1] had drawn special attention to these observations, Sir J. C. Ross, on his return from his Antarctic Expedition, considered that he had completely proved the truth of the theory that a temperature of 39°·5 was to be met with at different depths in different latitudes, and that this temperature presumably prevailed down to the bottom of the ocean.

The successive voyages of the 'Porcupine' and 'Challenger,' undertaken by the Admiralty at the suggestion of the Royal Society, set this question at rest, and showed that at great depths in the sea, in all latitudes where soundings were made, a temperature of about 32° prevailed.

As, therefore, the sea-water increases in density with a decrease in temperature, it is evident that the warmest water ought to be on top, and that the warm currents must overlie the cold ones. This is almost universally true, but in some well-authenticated instances a current of water warmer than that at the surface has been found flowing at some depth in the sea.

I shall, however, deal exclusively with the distribution of surface temperature. The first point to be noted is that the epochs of highest and lowest sea

[1] *Report of the Committee of Physics and Meteorology of the Royal Society*, 1840, p. 87.

temperature fall at least a month later than the corresponding turning points of the curve of air temperature. Speaking generally, these critical months for the sea are August and February, and as far as we can see, the latter of these months shows the most exceptional conditions of temperature distribution. It is easy to understand why this should be so, for a more extensive surface of the Arctic Ocean is covered with ice in February than in any other month, while at the same period the expanse of the southern oceans has been exposed to the action of solar heat throughout the southern summer, and is consequently at its warmest.

The charts which are reproduced in Plates IX. and X. are in general copied from those which were published by the Hydrographic Office in the 'Wind and Current Charts for the Pacific, Atlantic, and Indian Oceans,' 1872. As regards the Atlantic Ocean, the charts given in Mohn's 'Grundzüge der Meteorologie,' 1879, have been followed.

Commencing with the month of February (Plate IX.), we find a temperature of over 80° existing over the Western Pacific from 20° N. to 20° S., as far as the meridian of 180°. The area then becomes much narrower and, lying south of the Equator, disappears about the meridian of 130° W. A similar temperature prevails over the Indian Ocean down to 20° S., but in the Atlantic the corresponding warm area is much more contracted, forming a narrow belt along the equator from about 10° to 20° W., where it extends north and south to either tropic, on the coasts of the two Americas. This spreading out is the direct result of the bifurcation of the Atlantic equatorial current at Cape St. Roque, already mentioned.

The southern isotherm of 70° in the Pacific, follows approximately the parallel of 35° from the Australian coast for 140 degrees of longitude, as far as about 80° W. It there takes a sudden turn northwards through some fifteen degrees of latitude, under the influence of Humboldt's current. In the South Atlantic its course is somewhat similar, but the bend northwards on the meridian of Greenwich is not so sharp as that first described, and the curve turns southwards again, just touching the south point of Africa. From this it crosses the Indian Ocean in about the same latitude as it holds in the Pacific, but with a slight deviation northwards on the west coast of Australia.

The northern isotherm of 70° crosses the Pacific between the parallels of 20° and 30°, except in longitude 130° W., where it dips to the parallel of 18°, rising again before it strikes the American coast.

In the Atlantic its course is far more varied. It starts somewhat south of Cape Hatteras, dips slightly, and runs along the parallel of 28° to about 30° W., where it trends suddenly southwards and strikes the African coast near Sierra Leone.

The curves of 60° and 50° do not present much interest, as their course is straighter; but over the Atlantic the latter of these curves shows a very decided curvature to the northward, reaching a maximum in about 20° W. This feature becomes much more marked in the succeeding curves, which form decided loops, sharper and sharper as the latitude is higher. The apex of the isotherm of 40° lies near the Loffoden Islands, the curve thus passing over nearly thirty degrees of latitude and again descending to the centre of the North Sea.

The course of the line for 30° is still more erratic; starting from near Cape Race it passes north of Iceland, and when it approaches the meridian of Greenwich it begins to run due north. Its course is not drawn on the map beyond the parallel of 70°.

This extension of comparatively warm water into high latitudes in the North Atlantic is caused by the Gulf Stream pouring its volume of heated water along the channel which separates the New from the Old Continent, and bringing to the western shores of Europe the mild and damp climate which characterises them. The striking depression of the lower isotherms, which are packed close together along the coast of Nova Scotia and the New England States, is owing to the action of the Arctic current, already mentioned as forming the 'cold wall' of the Gulf Stream.

In the North Pacific the course of the isotherms for temperatures below 50° cannot be laid down with confidence, owing to the scarcity of material for such comparatively unfrequented waters. Almost the same thing may be said for the whole South Pacific Ocean. The only region exhibiting a temperature of 40°, which is habitually traversed by shipping, is that to the westward of Cape Horn, on the homeward track from Australia, and there the isotherm can be traced with some certainty. The same curve appears crossing the Southern Indian Ocean, but there its course is not well ascertained, owing to paucity of observations. *A fortiori*, the position of the curve for 30° must be almost imaginary, as, from the danger of ice, ships avoid the colder waters.

In August, Plate X., the region of 80° in the Western Pacific is displaced from its position in February

OCEAN CURRENTS AND SEA TEMPERATURE. 317

about ten degrees northward in the northern hemisphere and to a less extent in the southern. In the eastern half of the ocean, south of the Line, it stops short in about the same longitude as in February, but along the parallel of 10° N. a narrow belt stretches across the American coast. In the Indian Ocean this warm water does not reach the coast of Africa, as it disappears about ten degrees from it. In the Atlantic water of this temperature does not reach farther than to 40° W. in latitude 30° N. From this position it dips suddenly to 14° N. and, further to the eastward, to 10° N. It is entirely confined to the northern hemisphere, and the curve of 80° sweeps from the Cape Verde Islands round Bermuda, embracing the whole of the Caribbean Sea.

The curve of 70°, in the Indian Ocean, starts from the African coast, south of Natal, and keeps between 20° and 30°, when it turns sharply northward, almost to the Line, on the coast of Ecuador. In the North Pacific the corresponding curve, drawn northwards by the Kuro Siwo, is forced to dip through fifteen degrees towards the Equator, under the influence of colder water off the Californian coast. In the Atlantic the change is of the same character, but much less striking, as it only extends over some eight degrees of latitude.

The curve of 60° does not show any very remarkable features except on the eastern side of the Pacific, where it runs nearly parallel to the coast of California for some distance. In the South Pacific the corresponding curve runs nearly due northwards from 32° to 16° S. near the meridian of 80° W. In the South Atlantic the curve is deflected to the southward off the Cape of Good

Hope, under the influence of the warm Agulhas current from the equatorial regions.

In the North Atlantic the isotherm of 50° runs nearly due north-east, from the Banks of Newfoundland to the North Cape in Norway, just touching the south coast of Iceland *en route*. The same curve in the North Pacific has a very sinuous course, forming sharp loops near the south point of Kamtschatka, and rising to its highest latitude (about 56°) after passing the promontory of Alaska.

In all other parts of the sea the observations are so deficient in quantity that isotherms cannot be drawn with any confidence.

Both the charts show unmistakably the fact that in the southern hemisphere the general north-easterly set of the cold water brings the isotherms on the western coasts of the continents into lower latitudes, the action being most perceptible in the case of that for 60° on the coast of Peru. On the eastern coasts of Africa and South America, the warm waters of the equatorial currents exert an opposite action.

It is, however, in the temperate zone in the Atlantic Ocean that the greatest effect of ocean currents is made manifest, for there the curves show the sharp bends northwards which have been already described.

The form of the basin containing this ocean facilitates the warming action of the Gulf Stream, for it affords an easy passage into the Arctic Ocean for the water of that current, whereas in the Pacific the way for the Kuro Siwo is effectually barred by the narrowness and deficiency of depth of Behring's Straits. The water of this latter warm stream is therefore forced to return on itself at about the fiftieth parallel of latitude, and the Arctic Ocean, north of Behring's Straits, is left undisturbed by its heating influence.

The preceding brief sketch will give some slight general notion of the effects of currents in modifying sea-surface temperature; but to deal with the subject more fully and satisfactorily would pass beyond the limits of this work. It will, however, be interesting, before passing to another subject, to say a few words as to the sea temperature round our own coasts.

The Scottish Meteorological Society were the first in this field of inquiry, as continuous regular investigations into the temperature of the sea round Scotland were commenced by them more than twenty years ago. Isolated observations have also been made at a few spots on the coasts of England and Ireland, but no systematic series of observations on a uniform plan for the entire United Kingdom had been set on foot until the summer of 1879, when such were instituted by the Meteorological Council.

However, from some data existing in the Meteorological Office which were discussed by Mr. N. Whitley, and published, at least in part, by him in the Journal of the Royal Agricultural Society for 1868, the following general account may be given of the sea-surface temperature round our coasts in February and August, the epochs for which we have been considering the sea-surface temperature of the globe.

In February the temperature never reaches 50°. It closely approaches that degree at the entrance of the Channel, ranges about 45° all round the coast of Ireland, and falls nearly to 40° along the east coast of Great Britain, from the Thames to Shetland.

In August the change from the conditions just described is considerable. The sea round the Irish coast is about 10° warmer than in February, while along the English Channel, and as far up as the Wash, a uniform temperature of nearly 60° exists. In the last-named district the annual variation is nearly 20°. Passing farther north, we find stations giving a mean reading of nearly 55° on the North Sea coasts, and of fully that temperature on those exposed to the Atlantic. Here then the annual range is about 15°.

Of course these figures do not make any claim to precise accuracy, but they are sufficient to show that the mean temperature of the sea-surface is most constant at the entrance of the Channel, and least so off the mouth of the Thames. These are also the districts which are marked by great contrasts between the range of air temperature, which is small in Cornwall but comparatively large in the south-east of England, where the conditions most nearly approach those of the continent of Europe.

CHAPTER XVI.

THE DISTRIBUTION OF RAIN.

HAVING dealt with the subjects of the Temperature and Pressure of the Air, of the Winds, and of the Temperature and Currents of the Sea, the remaining factor which affects climate is the Distribution of Rain; and of this a brief account will be given.

The early rain charts of the globe exhibited certain belts, coloured in different tints, to show differences in the amount of fall. The average amount all over the globe was put down at about sixty inches, which was apportioned between the different geographical zones as follows:—The torrid zone was credited with 100 inches, the temperate zones with 30 inches, and the frigid zone with 15 inches. All such statements, however, rest on very questionable authority, for between the tropics the measured rainfall varies from 260 inches at Mahabuleshwur, on the Ghauts near Bombay, to two or three inches at Ascension, and to even less at some guano islands in the mid Pacific, to which I shall allude subsequently; while over the entire surface of the ocean there are no measurements at all that are of any value. As ships are in motion, we can only obtain from their logs entries of the number of times that rain was recorded, but not of its amount at any fixed station.

Y

The first advance on the simple belt rain map was to shade the coast-lines in certain places of a deeper tint; but if we compare a map on this principle, showing the total annual amount of rain, for example, that of Professor Loomis in the 'American Journal of Science and Art,' for January 1882, with the most recent map of seasonal rain distribution, that of Wojeikoff in vol. i. of the 'Zeitschrift für wissenschaftliche Geographie,' it is hard to realise that they represent only different aspects of the same phenomenon, the fall of rain.

It is impossible, in the present state of our knowledge, to give a true picture of these several elements, the amount, and the distribution in space and in time, of rainfall. If we were to attempt to draw monthly rain-charts we should at once be met by the difficulty that information for many portions of the globe is not attainable. The regions, in the northern hemisphere alone, in which practically no stations exist, are the interior and east of Asia; the whole of British North America, except the province of Canada; the entire Arctic Regions; and the Atlantic and Pacific Oceans. Curves could therefore be drawn with confidence over a very limited area. It is not with rainfall as with temperature or pressure, where corrections of more or less exactitude can be applied to reduce readings to their equivalents at sea-level. The distribution in vertical height of the amount of rain varies in different latitudes and with differences in the position of the mountain chains with regard to the prevailing wind, so that two stations situated close together may differ widely in the amount and in the seasonal distribution of their rain. Wojeikoff gives a very telling instance of this from two stations in Java, Batavia and Buitenzorg.

THE DISTRIBUTION OF RAIN. 323

The rainfall at the former place amounts to 81 inches, and the proportion between the quantity collected in the wettest month, January, and the driest, July, is 8 to 1. At Buitenzorg, situated only 25 miles from Batavia, and at an elevation of about 900 feet above it, the rainfall is 148 inches, and the proportion between the wettest month, March, and the driest, June, is only 2 to 1. The former station has true Monsoon rains, the latter has a fall which is nearly uniform in all the months. How then can the district of the island of Java, which contains these two stations, be fairly represented in any rain-chart of the globe? And yet in such charts, for many parts of the world, owing to the paucity of observing stations, the rainfall at one spot—a purely local phenomenon—has been assumed as a correct indication of the precipitation for an immense area.

I shall not, therefore, attempt any graphical representations of rain distribution, but shall give a brief description of its principal features.

In Chapter VIII. (p. 137), the three great agencies which are efficacious in bringing about precipitation of moisture have already been stated: ascending currents; the contact of warm air with the cool surface of the ground; and the mixture of masses of air of different temperatures. I shall now touch on the several factors which influence the distribution of the fall.

Of these the wind has to be considered first. Winds moving from high to low latitudes are generally dry, those moving in the opposite direction are generally damp. Winds blowing from off the shore on to the sea-surface, especially if the coast be bold and the land elevated, so that the wind is forced to descend, are essentially dry—the action being similar to that which gives its character to the *Föhn* of Switzerland—while winds blowing from the sea deposit their moisture on the coast. The South-east Trade wind, itself essentially dry at sea, becomes a rain-bringer to the mountains of eastern Brazil, and even to the eastern slopes of the Andes. For this reason, that the winds in low latitudes are Easterly, the eastern sides of tropical mountain ranges are, as Wojeikoff points out, often better watered and wooded than the western—the reverse of what is the experience in the temperate zone.

As already explained, this action is only, in a very slight degree, attributable to the contact of the warm air with the colder surface of the ground, for such an action can only affect directly the lowest strata of the atmosphere, unless very high mountains line the coast. The real cause is to be found in the enforced transference of the air into higher, and therefore colder, levels.

The equatorial zone of calms over the Atlantic and Pacific is a region where *it is alleged* that there are heavy and constant rains. These are due to the ascending movement of the air in the calm zone, as explained in Chapter XIII., which produces a constant canopy of dense cloud from the condensation of its moisture. This moisture is again restored to the earth in constant torrent-like showers, often accompanied by thunder and lightning. Caldcleugh's account of the regular occurrence of thunderstorms in Brazil has been already noticed (p. 146).

This statement of the perpetual rainfall near the equator may probably be accepted as a general assertion, but that the rain is not incessant in very low latitudes all round the globe is proved by the fall at Ascension

THE DISTRIBUTION OF RAIN. 325

(in 7° 55′ S.), already quoted, and also by the fact that in the Pacific, about 150° W., we have Malden Island, and some other islands, near to the equator (between the parallels of 6° N. and 11° S.), which are practically almost rainless, as is shown by their containing extensive guano deposits. Nay, more, as Wojeikoff shows, there is not a spot in the equatorial Atlantic Ocean where there are constant rains all the year round. There is always a contrast between the season when the Trade wind blows and that when the calms prevail. The latter are the Doldrums of Maury and the English writers, the 'Pot au Noir' of the French. In these regions, the rains follow the calms, and are essentially summer rains.

Although I have said that the rains follow the calms, it is not the case that they always fall when the sun is vertical. In the West Indies, as at Barbadoes, rains fall in all months, but chiefly in the autumn, the wettest month being October, two months after the sun has passed his second culmination.

The wettest districts of the globe are (1) parts of the calm zone above mentioned, and (2) certain localities where damp winds meet mountain ranges and are forced upwards. Such are, *par excellence*, the Khasia Hills in Assam, with Cherraponjee (where the rainfall is over 400 inches); the Western Ghauts with Mahabuleshwur, already quoted; the western coasts of the British Isles (Seathwaite); of Norway (Bergen); of North-west America (Sitka); of Southern Chili (Valdivia), and lastly, of New Zealand (Hokitika). All these latter places are exposed to Westerly winds blowing over an extensive tract of ocean and depositing their moisture on the first coast they meet.

The driest regions of the globe are, firstly, the great tract stretching eastwards from the Sahara through Arabia to Persia. This is continued through Central Asia over the southern Steppes of Russia to Lake Balkash. Farther eastwards again comes the desert of Gobi. This latter, however, owes its aridity in great measure to the fact that the mountains by which it is surrounded drain the wind of what little moisture it contains. In South Africa we have the great Kalahari desert; in North America the Great Salt Lake region; and in South America the narrow strip of Peru and Chili, between the Andes and the sea. The contrast between Santiago in 33° S. with a total rainfall of 14·10 inches, and Valdivia, only seven degrees of latitude further South, with a fall of 116·2 inches, is very remarkable. The interior of Australia, too, is rarely visited by rains.

From the estimates put down at the beginning of the chapter it will be seen that the frigid zones are put down as comparatively very dry regions. This is certainly true as regards the fall of rain, but the predominant mode of precipitation is in the form of snow, and among the characteristics of the Arctic Regions is the constant drifting of snow.

Nordenskiöld[1] says of it, at his winter station, Pitlekaj, near Behring's Strait: 'The fall of snow was not very great, but as there was in the course of the winter no thaw of such continuance that the snow was at any time covered with a coherent melted crust, a considerable portion of the snow that fell remained so loose that with the least puff of wind it was whirled

[1] *Voyage of the Vega*, translated by Leslie (London: Macmillans, 1881), vol. i., p. 483.

backwards and forwards. In a storm or strong breeze the snow was carried to higher strata of the atmosphere, which was so speedily filled with a close and fine snowdust that objects at the distance of a few metres could no longer be distinguished. But even when the wind was slight and the sky clear there ran a stream of snow some centimetres in height along the ground in the direction of the wind, and thus principally from N.W. to S.E. Even this shallow stream heaped up snowdrifts everywhere where there was any protection from the wind, and buried more certainly if less rapidly, than the drifting snow of the storm, exposed objects and trampled footpaths. The quantity of water which in a frozen form is removed in this certainly not deep, but uninterrupted and rapid, current over the north coast of Siberia to more southerly regions must be equal to the mass of water in the giant rivers of our globe, and play a sufficiently great rôle, among others as a carrier of cold to the most northerly forest regions, to receive the attention of meteorologists.'

It is very generally stated in text-books of physical geography that the Trade-wind regions are quite rainless. Maury expressed this positively in his 'Sailing Directions' (vol. i. p. 38, 8th edition): 'We know from observation that the Trade-wind regions of the ocean, beyond the immediate vicinity of the land, are for the most part rainless regions.' As regards the Atlantic and the Indian Oceans, however, the recent investigations of Köppen and Sprung[1] and of Von Danckelmann[2] respectively, have shown that the evidence

[1] *Annalen der Hydrographie*, 1880, p. 225.
[2] *Aus dem Archiv der deutschen Seewarte*, 1880.

obtainable from ships' logs as to the occurrence of showers, although, as already explained, it gives no information as to the amount of rain which falls, proves that the idea of any extended rainless region at sea is quite unfounded, and that the only districts in the Atlantic and Indian Oceans, respectively, where absolutely no rain falls are, in the former case, a small area on the eastern edge of the Trade-wind zone, and, in the latter, a limited region south of Java. In each instance the rainless interval occurs only in the spring of the respective hemispheres. It is probable that an investigation into the rainfall of the Pacific would lead to similar results.

Almost all of the rainless regions on land, however, owe their aridity to a similar cause to that cited for the desert of Gobi : viz. to the fact that they are shut off from the influence of moist winds by high chains of mountains. Dr. Hann points out a telling case of this in South America. In Peru, in the region of the Southeast Trade wind the coast is dry, but the eastern flanks of the Andes receive rain from the prevailing wind. Farther south, in the region of the Westerly winds, the continuation of the same range produces heavy precipitation at Valdivia, while the eastern coast of Patagonia, to leeward of the mountains, is very dry.

This action of mountains on the fall is conspicuous in many countries.

Hann[1] cites the Arlberg, in the Western Tyrol, where six stations are available. They are given in order from west to east, St. Christof being nearest the

---

[1] *Zeitschrift der österreichischen Gesellschaft für Meteorologie*, vol. xv., p. 373. Dr. Schncke, in the same volume at p. 498, gives a similar instance from the Black Forest.

top of the pass, and the two succeeding stations decidedly on the lee side of the mountain.

	Height in Feet	Rainfall (inches)
Bludenz	1,935	56·9
Klösterle	3,484	65·2
Stuben	4,609	82·2
St. Christof	5,899	86·2
St. Anton	4,255	39·1
Landeck	2,611	27·1

In general the amount of rain collected increases with the height above the sea, up to moderate elevations; but Hill[1] has shown that for the mountain stations of Hindostan the maximum fall occurs at about the height of 4,000 feet, being the level at which the South-west Monsoon is cooled just below its dew-point. Mahabuleshwur in the Ghauts, and Cherraponjee in the Khasia Hills, Assam, are both at about that level. Above that height the air is too cold to contain much vapour, and therefore cannot yield much rain, so that the amount decreases again as we ascend farther.

The law of the distribution of rainfall in vertical height is explained by Hill on the principle laid down by General Strachey ('Proceedings Royal Society,' 1861): that the quantity of vapour in the air at any level bears the same proportion to its quantity at the sea-level, as the maximum density of vapour for the temperature at the assumed level bears to its maximum density for the temperature prevailing at the sea-level There is no increase of vapour as we ascend.

In Europe the rainfall in the Austrian Alps has been carefully studied by Dr. Hann, and he finds that in parts of Central Europe he can recognise a level of maximum fall in winter at a height of from 3,000 to

[1] *Zeitschrift der ö. met. Gesellschaft*, vol. xiv. (1879) p., 161.

4,000 feet, but that in summer this level is far above the highest peaks. In the Alps, however, this fact could not be established, as the stations lay principally in valleys.

No rule can be stated with any confidence for the increase of fall with height in these islands; the precipitation is affected to such an extent by the trend of the valleys, so that the proportional difference is not the same in adjacent valleys on the different slopes of the same hill. Our own wettest district is that of the Cumberland lakes, especially where several valleys diverge from a common point; but recent discussions have shown that some of the Scotch glens, as Upper Glencoe, do not fall far short of their English neighbours.

The mere amount of water collected in the year does not enable us to compare districts satisfactorily as to their suitability for various crops. In one part of the country all the rain may fall in a wet season of two or three months' duration, and in another it may be distributed nearly equally over the whole year. Of course a protracted dry season checks vegetation for the time, and regions where these conditions prevail are, therefore, unsuited for pasturage. In the United Kingdom, especially in Ireland, a month without rain is a comparatively rare phenomenon, and, for grazing purposes, the grass hardly ever fails in summer.

It has already been stated that the rains of low latitudes are essentially summer rains, as they occur principally when the sun is highest. In Mexico the dry season is called *verano*, summer; the wet season *invierno*, winter, although then the sun is highest. In the West Indies the year, in popular language, is divided according to the periodical recurrence of the rains—the *seasons*, as they are called.

Wherever the Trade winds blow there is comparatively little rain, unless where the wind blows on to a mountainous coast. With the periodical shifting of the Trade-wind areas the dry areas shift likewise; and the descent of the Return Trade, which, as we have seen, is the overflow from the calm belt, brings with it abundance of rain. The district where this occurs, extending approximately from lat. 30° to 40° in both hemispheres, receives its rain when the sun is lowest. This is, therefore, the region of the winter rains, and the entire subtropical area is characterised by the same features. This region embraces the countries bordering the Mediterranean, with Asia Minor and the western part of Persia. In North America, California and Oregon belong to it, and in the southern hemisphere the Cape Colony, South-west Australia, and the northern island of New Zealand.

There are, however, certain districts lying between the parallels of latitude above given which have summer and not winter rains, and which, therefore, do not come into the subtropical rain district. These are the eastern coasts of the great continents, as China and the Eastern States of the Union, which enjoy a sort of monsoon rain in the height of summer. Natal, in Africa, and the Argentine Republic come under the same category. As all these countries receive abundant rains at the period most important for the growth of crops they are most favourably circumstanced as regards agricultural industry. The countries with winter rains and summer drought must have recourse to irrigation to water their fields.

To the subtropical region succeeds the region of rain at all seasons, and to this the British Isles belong.

Here the rainfall depends on the somewhat irregular succession of barometrical depressions and anti-cyclones which are constantly moving over the earth's surface in the temperate zone, and which, as we shall see in the subsequent chapters, determine our weather. As a general rule we may say that the western coasts of the continents have autumn rains, gradually passing into summer rains as we advance into the interior of the country.

In these islands, on the west coast, the wettest month is January, thus bearing out the general sub-tropical character of the climate of our Atlantic coast, but in October we find a second maximum. On the east coast this latter becomes the chief maximum, so that we have true autumn rains; but the difference between the several months is not great in London, the wettest month, October, giving 2·74 inches, on an average of sixty years, while the driest, February, gives as much as 1·50 inches.

The following are the figures—firstly, for the sixty years, 1813-72, as determined by Mr. Dines,[1] and, secondly, for the fifteen years, 1865-80, as calculated from the observations at Kew Observatory, recently published by the Meteorological Office:[2]—

	1813-72	1865-80		1813-72	1865-80
January	1·91	2·16	July	2·32	2·42
February	1·50	1·71	August	2·26	2·20
March	1·52	1·31	September	2·35	2·50
April	1·66	1·85	October	2·74	2·48
May	2·07	1·63	November	2·28	1·90
June	2·01	2·13	December	1·93	2·23

The curve for the shorter period is naturally far less regular than that for the longer (fig. 38), and it will

---

[1] *Quarterly Weather Report*, 1873. Appendix, p. [13.]
[2] Rainfall Tables of the United Kingdom, 1882.

be noticed that the secondary maximum in December which appears in the second column is entirely absent from the first.

Dr. Hann, in the work which has supplied most of the materials for this chapter, gives an account of the monthly distribution of rainfall in Europe, which would be too long for incorporation here. Among the most interesting facts which it exhibits is that the Alps divide the region of summer rains of northern, from

FIG. 38.

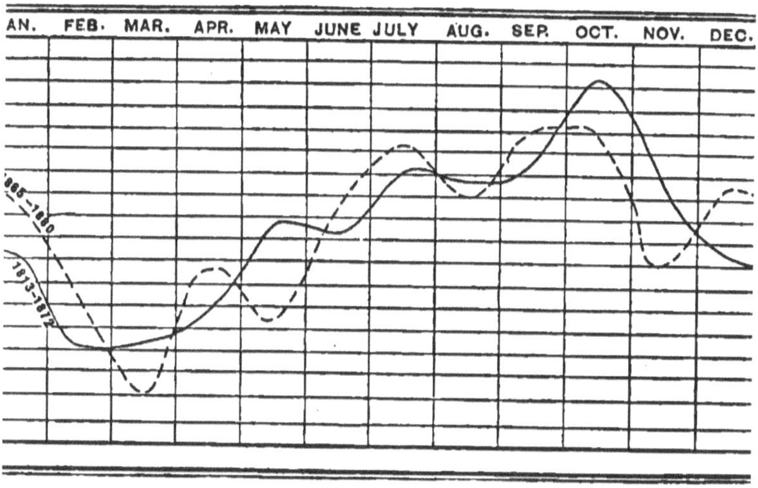

Monthly Rainfall in London.

that of the autumn rains of southern, Europe. The following figures, which are percentages of the total annual fall, show the gradual increase of the predominance of summer rains as we pass eastwards over Northern Europe. In North-west France 24 per cent. of the annual fall occurs in summer; in Northern Germany the percentage is 28; in Northern Prussia, 36; in Central Russia, 38; and in the Ural district, 53; so that in the last-named district more than half the rain falls in summer.

Rain Probability.—If we divide the number of days in a month on which a measurable quantity of rain (at least 0·01 inch) has been recorded, by the total number of days in the month, we get a quotient which represents the mean probability that rain will fall on any day in that month. The subjoined table shows the differences, as regards this probability, for 0·01 inch between Valencia in the south-west of Ireland, with a fall of 59·9 inches in 235 days, and London with a fall of 25·5 inches in 168 days. The period is that of the five years 1871-5. To these are added the probabilities, for 0·01 inch and for 0·1 inch respectively, for Cobham in Surrey, for the forty years, 1826-65 inclusive, which have been supplied to me by Mr. G. Dines. The figures are represented on the diagram fig. 39, p. 335:

Monthly Rain Probability.

	Valencia—1871-5		London—1871-5		Cobham—1826-65	
	No. of Days	Probability	No. of Days	Probability	Probability	Probability
	0·01 Inch	0·01 Inch	0·01 Inch	0·01 Inch	0·01 Inch	0·1 Inch
Jan. .	28	0·90	19	0·61	0·44	0·18
Feb. .	20	0·70	14	0·50	0·45	0·16
March	18	0·58	13	0·42	0·43	0·16
April.	15	0·50	13	0·43	0·42	0·18
May .	13	0·42	11	0·35	0·42	0·19
June .	16	0·53	13	0·43	0·44	0·19
July .	20	0·64	14	0·45	0·42	0·19
Aug. .	19	0·61	12	0·39	0·41	0·20
Sept..	19	0·63	13	0·43	0·47	0·23
Oct. .	25	0·81	17	0·55	0·49	0·25
Nov. .	19	0·63	16	0·52	0·50	0·22
Dec. .	23	0·74	13	0·42	0·46	0·18

This shows us that at Valencia, in January, out of ten days the chance is that only one will be perfectly dry, while in London, in the same month, we may hope for four dry days out of ten. The only month at

Valencia where there is a slightly better chance of dry
than of wet weather is May, which the foregoing
table, shows to be also the driest month in London.

The figures for Cobham, however, are very interesting as compared with those for London. In the first
place, the predominance of January, as the wettest
month, does not hold for the longer period at Cobham,
where the highest probability is for November and
the lowest for August, instead of May. These figures

FIG. 89.

[Figure: Monthly rainfall probability curves for Valencia 1871-1875, London 1871-1875, and Cobham 1826-1865, with months January through December on the horizontal axis.]

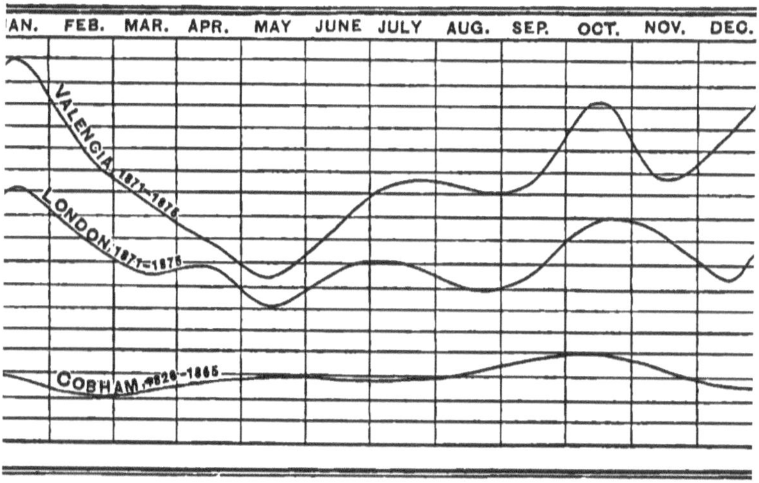

Monthly Rain Probability for 0·01 in. for Valencia and London, for
0·1 in. for Cobham.

are, however, only for a hundredth of an inch of rain,
an amount which is practically insignificant. The last
column shows for Cobham the probability of the fall of
a tenth of an inch of rain, that is, *of what might ordinarily be called a wet day*. This shows a very decided
maximum in the autumn months and a minimum in
February and March, months which are shown to be
nearly as damp as any, as far as the absolute occurrence of any rain is recorded. These facts show us

that in the winter small rain and mist contribute largely to make up the total fall, while in the autumn the rain comes down in heavy showers.

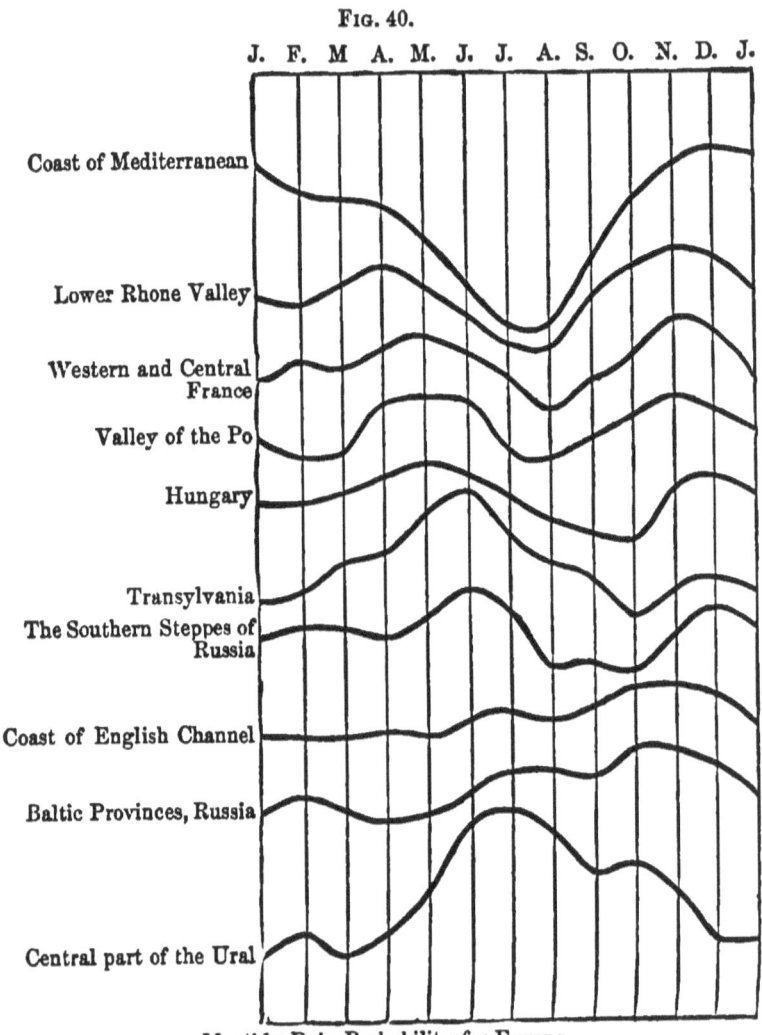

Fig. 40.

Monthly Rain Probability for Europe.

Dr. Köppen[1] gives the preceding diagram (fig. 40) of the rain probability over Europe, which shows very clearly

[1] 'Ueber Regen wahrscheinlichkeit in einigen Theilen Europas,' *Zeitschrift der ö. met. Gesellschaft*, vol. iii. 1868, p. 497.

where we should not go if we want fine weather in our autumn holidays. On the Mediterranean coast the curve is at its lowest in summer. In Hungary the lowest point is in October and the highest in May. In Transylvania the same features are more strongly marked, but the wettest month is June. The conditions for the Channel coast and the Baltic are generally similar to those shown in the table on p. 334 for the British Isles, while the predominance of summer rains in the Urals is very decided.

CHAPTER XVII.

CLIMATE.

UNDER the term Climate we understand the combined effect of all the phenomena whose distribution over the globe we have recently been considering, and which determine the suitability of various districts for the support of their respective fauna and flora. Beyond all doubt the most important factor in these determinations is the temperature. We all know that vegetation attains its greatest luxuriance in the damp heat of the tropical regions, while it can hardly be said to exist in the extreme cold of the polar lands. The law of decrement of heat with ascent above the sea-level has already been explained, and, accordingly, it will be understood, as Herschel says,[1] 'that in ascending a mountain from the sea-level to the limit of perpetual snow, we pass through the same series of climates, so far as temperature is concerned, which we should do by travelling from the same station to the polar regions of the globe; and in a country where very great differences of level exist we find every variety of climate arranged in zones according to the altitude, and characterised by the vegetable productions appropriate to their habitual temperatures.'

The old division of the earth by Parmenides into

[1] *Physical Geography*, Section 249.

the five zones of geography has been found to be quite inadequate as a representation of the climatology of the globe. Merely to take the phenomenon which we have most recently been treating, the distribution of rainfall, we have seen that the subtropical conditions of fall extend far up into the temperate zone. We are, however, hardly prepared to agree with Professor Supan, who, in Petermann's 'Mittheilungen' for 1879, proposed a formal abandonment of the old zones in favour of a system more in accordance with climatic facts.

Speaking very generally, we may say that the distribution of the plants most important to mankind, such as the cereals, depends on the summer temperature, for the temperature prevailing in winter, when the seed is not in the ground at all, exercises only an indirect effect by its influence on the condition of the soil. The district of Manitoba yields magnificent crops of wheat, despite its winter temperature, which often falls far below zero of the Fahrenheit scale.

The distribution of animals is more dependent on the winter temperature, and, as animals can bear a greater range of temperature than plants, their distribution is more extensive.

The contrast between different localities as to the extent of variation of temperature which they experience in the yearly period is a factor of great importance in the determination of climate, and one to which, as yet, comparatively little attention has been directed.

The first to draw Range Maps of the globe was the late A. Keith Johnston, jun., who, in the 'Proceedings of the Royal Society of Edinburgh' for 1869, published such maps, but on a very small scale, and on the polar projection. The maps were reproduced in his 'Physi

Geography,' 1870. Lately the subject has been taken up independently by Professor Supan, whose paper appeared in the first volume of Kettler's 'Zeitschrift für wissenschaftliche Geographie,' and whose map is reproduced in Plate XI.—at least, in so far as the difference between the Fahrenheit and Celsius scale permits us to do so.

It will be seen from it that the regions of very great annual range are all situated within the land areas of the northern hemisphere. In fact, near Jakutsk, in Siberia, we find a small district with the enormous range of 100° F.; Jakutsk itself having a temperature of 65°·8 in its warmest and −44°·9 in its coldest month These figures even fall short of those for Werchojansk. On the American side the extremes are not so great, but yet Northumberland Sound has a range of over 75°, from 36°·7 to − 38°·6, but it must be observed that this statement is founded on only one year's observations.

The curve for 60° range, in Asia, passes from the Straits of Kara southwards to the Caspian, thence eastwards to the head of the Gulf of Pechili, whence it follows the coast pretty closely to near Petropaulowsk in Kamtschatka. In America it runs south-eastwards from Behring's Straits to the western end of Lake Superior, and thence northwards again till it strikes the Greenland coast, near Upernavik.

The line of 40° encloses some of the most fertile tracts of the globe, as will be seen by its embracing most of the United States. In the eastern hemisphere it follows an irregular course from the North Cape to Central Arabia, being sharply diverted from the direct line on crossing the Gulf of Finland, and again as it passes the Black Sea. It runs to the northward of the Persian Gulf, and thence nearly due east to the south point of Japan.

The curve for 20° first skirts the west coast of America as far as Mexico, crosses Florida, and, trending north-eastwards to the Faroes, turns sharply, and runs nearly due south along the west of the United Kingdom and of Portugal to North Africa, where it passes nearly to the Gold Coast, and from thence it skirts the north coast of the Indian Ocean to the south of China.

The same curve is shown south of the equator in three distinct areas, in Southern Brazil, in South Africa, and in Australia. This is the extreme range known certainly in the southern hemisphere, for the insertion by Dr. Supan of the curve for 20° over the Southern Ocean seems very doubtful, owing to the deficiency of information existing for that part of the world.

The rest of the globe, embracing by far the greater portion of its area, enjoys a very uniform temperature throughout the year, the range nowhere exceeding 20° F. This region extends nearly up to the sixtieth parallel on the west coast of North America, and beyond it in the neighbourhood of the Faroe Islands in the Atlantic. Almost the whole of the sea-surface of the globe belongs to it, as well as the major portions of South America and South Africa.

The principal laws of distribution of annual range of temperature, as given by Dr. Supan, may be thus summarised:—

1. The range increases from the equator towards the poles, and from the coast towards the interior of a continent.

2. The regions of extreme range in the northern hemisphere coincide approximately with the district of lowest temperature in winter. On the whole the range curves in their course resemble the isotherms of January.

3. The range is greater in the northern than in the southern hemisphere.

4. In the middle and higher latitudes of both hemispheres, with the exception of Greenland and Patagonia the western coasts have a less range than the eastern.

5. In the interior of the continents the range, in mountainous districts, diminishes with the height above the sea.

The influence of the sea in moderating the extremes of climate is unmistakable on inspection of the map but even more decidedly do the agencies of prevailing winds and prevailing currents show their effects Where the prevailing winds are Westerly, the so-called Anti-trades, and where the ocean currents are flowing from the equator into higher latitudes, the cold of winter is mitigated, and the curves of equal range bent polewards. On the other hand, on the eastern coast of America, and even more so on that of Asia the prevailing winds are Northerly and cold, so that the temperature in winter falls very low, while the summer is comparatively warm, and the contrast between the opposite seasons is very marked.

It has already been said that the distribution of animals and plants is in great measure regulated by conditions of climate, and text-books on physical geography give abundant illustrations of the truth of this statement. The individual genera are in great measure confined to, or at all events are met with in greatest perfection in, the regions which afford certain conditions of climate. These conditions, however, involve, over and above the mere temperature, the amount of sunshine which falls on the district, and the supply of moisture both as to its total amount and as to the seasonal distribution of its fall.

The general mildness and equability in high northern latitudes of the climates on the western coasts of continents, as compared with the eastern, of which the isothermal, the isabnormal, and the range charts offer such striking evidence, is, as has already been said, mainly attributable to the general prevalence of Westerly winds, with the moisture they bring with them, and to the set of warm ocean currents on their shores.

The exception to this rule, in the case of Greenland, which has already been noticed by Dr. Supan, is due to the fact that the eastern coast of that island is swept by the cold Arctic current which we have already mentioned as affecting the climate of the eastern coast of North America, and which, owing to the immense mass of ice it brings with it, at all seasons, effectually prevents the temperature in summer from rising to any material extent, and so reduces the annual range to an amount below that which obtains on the inhabited west coast of Greenland.

In the southern hemisphere the same difference, between the west and east coasts, is not so perceptible in high latitudes, for there the continents do not extend far into the temperate zone, with the exception of South America, which at its southern extremity is so attenuated that there is no space to develop any great difference in climatic conditions between its eastern and western shores.

In fact, the tendency is in the opposite direction, for the western coast of Patagonia is somewhat colder than the eastern. This is due to the action of Humboldt's current, bringing up the cold water of the Southern Ocean. This influence of the set of water from the south pole is most perceptible in lower latitudes, especially between the parallels of 20° and 30°, where, on the western coasts of both South America and Africa, the cold drift from the Antarctic Ocean, mentioned at p. 293, exerts its full effect.

However, even independently of its motion, the mere presence of water plays a great part in modifying climate. This is, above all, noticeable in the climates of islands, as compared with the adjacent mainland. Hence comes the subdivision of climates into *insular* or *moderate*, and *continental* or *excessive*. The west coasts of continents enjoy insular, while the east coasts and the interior have to put up with continental, climates.

If we pass from the subject of climate, considered generally, to that of the climate of particular stations or districts, we find that this is regulated mainly by the influence of the wind. If a station is sheltered from cold winds by a range of hills, a cliff, or even a belt of wood, while at the same time it is open to warm winds and has a southern aspect, we have the principal conditions of a winter health-resort. The character of the soil is also a most important factor in the salubrity of a locality. If the soil is light and sandy, so as to allow of the rapid removal from the surface of the rain water which falls on it, the climate is far drier than can ever be the case in close clayey soils.

The south coast of England presents us with many situations which afford great advantages for invalids;

the most important differences between these consist in the degree of shelter, from occasional cold winds, which can be secured at each.

Again, the climate varies greatly with the elevation of the station. Not only, on the one hand, does the mean temperature fall with height above sea-level, but, on the other, a position on a slight eminence or on the side of a hill secures the residents immunity from the extreme severity of frost, inasmuch as in calm weather the cold air gradually sinks down to the lowest level of the valley. This is a phenomenon akin to the 'upbank thaw' already mentioned, p. 213. It is a well-known fact in pleasure-grounds that evergreens in low-lying bottoms suffer far more in winter than those growing on the slopes above. Moreover, low-lying stations are also notoriously far more liable to the visitations of fog than the uplands above.

Some localities are specially favoured, even though they do not lie on a south coast, as for instance the coasts of the Moray Firth, Nairnshire, and the Carse of Sutherland. These districts owe their good fortune mainly to the fact of their lying on the lee-side of an extensive mountain district. The prevailing winds, being Westerly, deposit their moisture on the Atlantic slopes of the hills, and the air, thus drained of its moisture by ascent, is further dried and warmed by its descent to the sea-level on crossing the island. The atmosphere of these districts is therefore essentially drier than that of other places in the same latitude, and the dryness, inducing clearness of the sky, allows more frequent access to the sun's rays than is possible on the cloudy west coast. Accordingly crops will ripen in the valley of the Shin, in Sutherlandshire, in 58° N.

which never come to maturity in Argyllshire, two degrees farther south.

The climate of hill stations in tropical countries is also especially enjoyable and salubrious, as compared with that of the plains. This is almost entirely due to the reduction of temperature with height. At the elevation of, say, 6,000 feet, the mean temperature will be reduced, in round numbers, 20°, so that where there is a temperature of 80° on the plains there will be one of 60° above.

Another particular in which high-level stations differ from those below them is the advantage they enjoy in the receipt of a greater amount of direct heat from the sun, owing to the fact already mentioned that the densest and dampest strata of the atmosphere lie the lowest.

It is mainly to this dryness of the atmosphere, and consequent intensity of solar heat, that the favourite Swiss station of Davos, at the level of 5,000 feet, owes its recently-acquired popularity as a winter residence. In some measure, however, it derives the comparative mildness of its climate from its situation in a valley sheltered from all except the warm South-west winds.

The increasing dryness of the atmosphere with elevation may, however, have its serious disadvantages; for when we come to high table lands, like those of Central Asia, we find insufficient moisture in the air to check the radiation to the earth by day or from the earth by night, so that here the range of temperature in the twenty-four hours is often more than the strongest constitution can bear.

It would, however, far exceed the limits at my disposal were I to discuss local climate, and the causes which affect it, at the length which the importance of the subject demands.

For such a discussion the reader must be referred to such works as the 'Physical Geography' of Herschel or of Mrs. Somerville, while as regards the question of climate in its relation to health and disease, he will find full information in numerous works of deservedly high reputation on medical climatology.

CHAPTER XVIII.

WEATHER.

WEATHER may be described as the combined effect of all the different meteorological conditions which have been treated of in the foregoing pages.

In popular phraseology the term Weather has more especial reference to rain and temperature than to any other phenomena, but it will be explained how all the meteorological conditions are linked together, and how we cannot consider the rain without reference to the wind which brings it, or the wind without reference to the distribution of barometrical pressure to which it is due. The relations of temperature to the other conditions of weather have already come before us frequently.

Prior to the introduction of the modern system of weather observations, taken simultaneously over an extensive area, and telegraphed to some central office where they are inserted on maps, so as to give a picture of the weather existing at the time of observation at each station belonging to the system, observers were compelled to discuss the probabilities of weather for their own locality by the help of the records of observations taken in that locality alone.

This necessity led to the calculation of what may be called 'weather windroses'; exhibiting the average values of pressure, temperature, humidity, &c., &c., for the wind from every point of the compass. This work has been carried out in detail for a large number of stations in different parts of the world. The calculation is laborious but simple, and the form of the windroses may be varied in many ways. An obvious modification, from which Professor Dove deduced his practical rules for weather, was the determination of the change in the readings of each instrument in passing from one point of the compass to the next.

The following is a condensed specimen of Dove's first two rules:—

The $\begin{Bmatrix} \text{barometer falls} \\ \text{thermometer rises} \end{Bmatrix}$ with E., S.E., and S. winds; with a S.W. wind it ceases to $\begin{Bmatrix} \text{fall} \\ \text{rise} \end{Bmatrix}$ and begins to $\begin{Bmatrix} \text{rise} \\ \text{fall} \end{Bmatrix}$; it $\begin{Bmatrix} \text{rises} \\ \text{falls} \end{Bmatrix}$ with W., N.W., and N. winds; with a N.E. wind it ceases to $\begin{Bmatrix} \text{rise} \\ \text{fall} \end{Bmatrix}$, and begins to $\begin{Bmatrix} \text{fall} \\ \text{rise} \end{Bmatrix}$

Dove propounded these and other similar rules as applicable to the entire northern hemisphere, and gave an analogous set, *mutatis mutandis*, for the southern hemisphere, where, of course, a Southerly wind, blowing from the pole towards the equator, corresponds to a Northerly wind in our latitudes.

A great misconception, if one can apply the word to a man of such eminence as Dove, underlying these propositions, consisted in the assumption that what held good for Western Europe was true for the entire northern hemisphere.

Meteorology, and more especially the study of weather changes, has suffered seriously by having been studied too exclusively with regard to local experiences in Western Europe. The calculated windroses show undeniably that for Germany the barometer does stand highest, and the thermometer lowest, with N.E. winds, but, as has already been pointed out in Chapter XIV., p. 278, along the north coast of Europe and Asia the coldest point of the windrose shifts through 108° from N. 62° E. (nearly E.N.E.) over the North Sea to N. 46° W. (N.W.) in Eastern Asia. In every case the coldest wind comes from the direction in which the region of the lowest temperature for the season lies. In Germany, in winter, this is situated to the northeastward, over Russia; for Eastern Asia it lies near Jakutsk, near the Siberian pole of cold.

Again, as regards pressure, although it is a fact that in Germany the barometer stands highest with N.E. and lowest with S.W. winds, in Greenland it stands equally high with all winds.

That the same values for the windrose would not suit all seasons was fully known to Dove, who pointed out repeatedly that the changes in the meteorological instruments were more decided in winter than in summer, and, as we have already explained in Chapter XII., p. 233, it is in winter that the strongest contrast in conditions of temperature is found.

The most exhaustive calculation of windroses that has ever been carried out is that of P. Schreiber for Leipzig ('Petermann's Mittheilungen,' Ergänzungsheft 66), who has arrived at the following rules:—

1. There is no relation, in the first instance, between barometrical pressure and wind-direction, both

being functions of the general conditions of weather over large areas.

2. Temperature depends in the first instance on wind-direction; in the second on the pressure, for the higher the barometer the lower is the temperature of all winds. In summer alone is there an exception to this rule. When the barometer is very low all winds have a warming action.

3. The action of wind-direction and barometrical pressure on vapour tension is very complicated, and is of slight importance. In general the warmest currents are the dampest.

4. Relative humidity, cloud, and the probability, as well as the amount, of rain, are in the first instance functions of pressure, and are, roughly speaking, in an inverse ratio to it. Nevertheless they have also a secondary relation to wind-direction.

Now, as Dr. Schreiber points out, these results justify the application of the term 'weather-glass' pre-eminently to the barometer, but they also show how many factors must be taken into consideration if we wish to form a judgment on coming weather by means of instrumental observations.

The fact is, as just stated, that local barometrical pressure and wind-direction are both functions of the general conditions of weather over large areas, and what these conditions are I shall endeavour briefly to explain.

If we lay down on a map a number of barometrical readings taken at the same time over an extended area, we find that the readings are never the same all over the space, and that in their variations they exhibit a methodical arrangement. In one part of the map the barometer will stand high and in another low. We shall not find an isolated high reading surrounded by a number of low ones, or, *vice versâ*, a single low barometer in a part of the country where the other barometers are high, but each district, high or low, will show a gradual progression towards the relatively highest or lowest reading within its area.

In fact, if we suppose the atmosphere to be a liquid, like the sea, and enveloping the earth, we may imagine that portion of it which lies over the Atlantic Ocean in our latitudes to be like a gigantic river, for it has already been explained (p. 274) that the general motion of the air over us is from the westward. On the surface of such a river we often see small waves and eddies, each with its own circulation, which are carried along with the stream.

If we could look at the upper surface of the atmosphere, *supposing it to be homogeneous and of equal density throughout*, we should see somewhat similar conditions, what corresponds to the crest of the waves being patches of excessive pressure, while what corresponds to the eddies would be the areas of defective pressure. The surface, if suddenly congealed, would present the appearance of a number of isolated mounds and pits, with very little level area at all.

The contour lines on the surface, to use a geographical expression, would be the lines of equal barometrical pressure, or the 'isobars,' as they are called, which are shown on all weather maps.

The pits would have steep sides, be deep, and of comparatively small diameter; while the mounds, on the contrary, would be flat-topped and extensive in area. The pits are what are called barometrical depressions or cyclonic systems, from the Greek word κύκλος, a circle, because the wind whirls round them; the mounds are the areas of high pressure, called 'anti-cyclones,' because they are the opposite of cyclones.

If the reader has got well into his mind this fact of pressure being unequal over a large area, the next things for him to remember are that the greater the inequality of pressure over the area, the stronger is the wind, and that the wind-direction is also determined by the distribution of the pressure.

This relation of the wind, both in direction and force, to the distribution of pressure, instead of to the actual reading of the barometer at each place, is the essential feature of modern meteorology, and a moment's reflection will show that it greatly diminishes the value of all the rules which are given for predicting weather by the readings of a single barometer. We require to know not only what our own barometer tells us, but also how our neighbours' instruments, in all the country round, differ from it. This latter knowledge is, of course, only obtainable by telegraphic means.

The principle of the relation of wind to differences of pressure, known as Buys Ballot's Law, was explained in Chapter XIII. (p. 255).

When we say that the force of the wind depends on differences of pressure, we must not omit to mention that meteorologists have agreed to express these differences of pressure in terms of what are called 'gradients.' Gradients are the differences of pressure over a given distance. The units selected, by international agreement, are respectively 0·01 in. for pressure, and fifteen geographical miles for distance (corresponding to one millimetre and one degree [sixty geographical miles]). Further explanations of the calculation and use of gradients will be found in my 'Weather Charts and Storm Warnings,'[1] but it may here be said that no simple relation between the gradient and the force of the wind has yet been established. The difficulty experienced in determining this relation arises in part from the fact that the wind does not always blow horizontally, while the gradients are calculated for sea-level pressures, and therefore for the horizontal plane. Local irregularities in the surface of the ground also seriously affect the force as well as the direction of the wind.

If we apply Buys Ballot's Law to our areas of high and low barometer, we see that round the former, the anti-cyclones, the air must revolve *with* watch-hands; that if we were in a balloon, drifting before the wind, and always kept the highest barometer on our right-hand side, we should travel from east to west, round by south, keeping on the southern edge of the anti-cyclone. In the depressions, on the contrary, the balloonist, always keeping the point of lowest barometer on his left-hand side, would move from west to east, round by south, keeping on the southern edge of the system. This would be a motion *against* watch-hands.

This will appear from the accompanying diagrams, of which fig. 41 represents a cyclonic system, accompanied by a very serious storm, which raged in these islands on Sunday, November 29, 1874. The lines drawn on the map are isobars, which, in this case, are

[1] *Weather Charts and Storm Warnings*, by Robert H. Scott, F.R.S. H. S. King & Co, 1876.

hown for every two-tenths of an inch, the central area eing bounded by the isobar of 28·6 inches, which

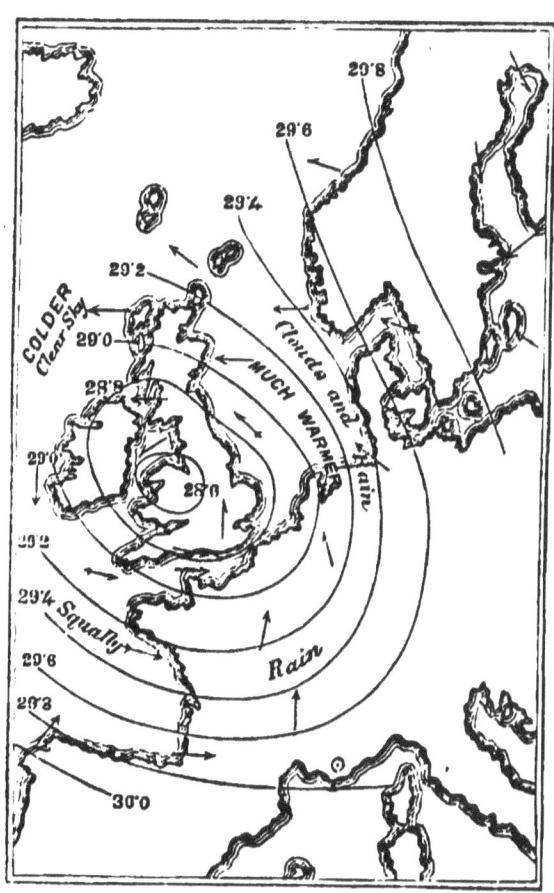

Fig. 41.

Weather Chart, Nov. 29, 1874, 8 A.M., Cyclonic System.

The arrows are thus distinguished, according to Beaufort's scale, p. 159:—
    ➼ forces above 10.   ↠ forces 8 to 10.
    → ,, 5 to 7.   ⇁ ,, 1 to 4.
        ⊙ dead calm.

ιcloses North Wales, Cheshire, and part of Lancashire. The wind is shown by arrows which fly with it, and

its force is indicated by the number of barbs or feathe
on the arrow. For instance, the arrows for Scilly, tl
Bay of Biscay, and Scarborough denote very hea
gales, from West-north-west in the two first, and fro

Fig. 42.

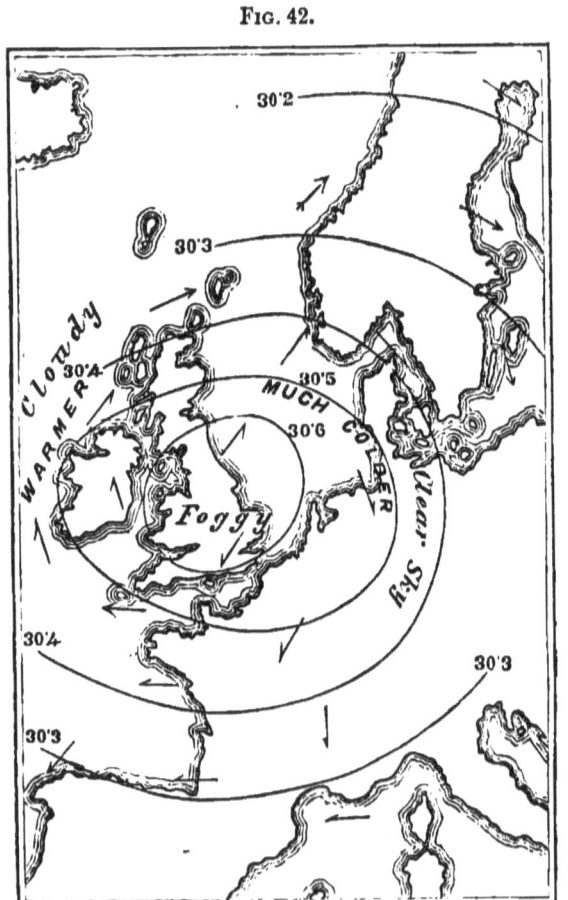

Weather Chart, Feb. 4, 1874, 8 A.M., Anti-Cyclonic System.
For explanation of the wind arrows see fig. 41, p. 355.

South-east in the last case. The weather, temperatur
and the state of the sky are given in words.

Fig. 42 represents an anti-cyclonic system, as

existed on February 4, 1874, a day forming one of a long series exhibiting similar conditions of weather, which prevailed in that month.

The isobars are drawn for every tenth of an inch, and the remaining indications on the map correspond to those explained above.

These diagrams show how it is that, knowing the distribution of pressure, we can tell the direction of the wind, and *vice versâ*.

The idea which meteorologists have of these two classes of areas is that the lower air is flowing on a spiral course *into* the areas of depression, where it rises from the surface of the earth, while it is flowing on a spiral course *out of* the areas of high pressure, where it descends from the upper regions of the atmosphere.

The systems differ essentially in the character of the weather which accompanies them. As a rule, depressions are associated with a cloudy sky, dampness of the air, and rain. These conditions imply warmth in winter and cold weather in summer. Anti-cyclones, on the contrary, are accompanied by a dry atmosphere, and a sky generally clear of clouds, though fogs are very prevalent. These conditions produce cold in winter, and heat in summer.

Again, in depressions the isobars lie close together, and the winds are strong. In anti-cyclones the isobars lie far apart, and the winds are light. The instances represented in figs. 41 and 42 illustrate these facts very clearly, more especially when it is remembered that the isobars are drawn for every tenth of an inch in the latter figure, but only for every two-tenths in the former.

The statements which have just been made are of

course of a very general nature, but they furnish the key to most of the phases of weather which pass over us.

Let us now consider, in a little more detail, the weather phenomena of typical depressions and anti-cyclones. The diagrams have shown us the wind conditions in each case. The map for November 29 (fig. 41, p. 355) exhibits the wind circulating according to Buys Ballot's law, against watch-hands, and blowing hard; in fact a very heavy gale. As regards temperature, the whole eastern part, or front, of the system which was advancing from west to east, was marked by a very rapid rise of the thermometer, the weather was very warm for the season; while on the western side or in the rear of the system, the temperature had fallen, and the weather was cold. In both these cases the terms refer to the change of temperature during the preceding twenty-four hours. Where the weather was warm there was a great deal of cloud and rain; where it was cold the sky was clear.

The anti-cyclone (fig. 42, p. 356) presents us with conditions in exact contrast to those just enumerated. On its eastern side the weather had become much colder since the previous day, and on its western side the temperature had begun to rise. The sky was generally clear, though foggy at the centre of the system, but clouds showed themselves on our western coasts, while no rain was reported anywhere.

I shall now describe very briefly our ordinary experience in the British Isles, when one of these depressions passes over us in winter, say after a frost. The sky first covers itself with thin mare's-tail clouds which gradually become denser, and develop themselves into parallel bars, stretching across the heavens, and which, owing to perspective, seem to converge at opposite points of the horizon, forming the appearance of the Noah's Ark (Chapter IX.). This arrangement, under the circumstances described, stretches from south-east towards north-west, and the wind at the surface of the ground soon shifts to a South-easterly direction, while the barometer begins to fall rapidly, the thermometer to rise, and fine rain sets in. The changes progress steadily, the barometer falls more and more, the wind shifts through South to South-west, and blows harder the more rapidly the mercury sinks in the barometer-tube, the cloud canopy grows denser and denser, and the rain, first setting in as mist, becomes heavier and more constant. At last, when the barometer has reached its lowest point, the wind flies round to West, or West-north-west, in a very heavy shower; the instant of shift of wind being marked by a sudden rise of the barometer and an equally sudden fall of the thermometer. The sky soon clears, and cumulus clouds drift past, producing occasional heavy showers, which are often accompanied by thunder and lightning with hail, the wind still drawing more and more towards North and the barometer rising rapidly.

This is a general description of the changes of weather when the observer is situated on the southern side of the track of the centre of the cyclone; and this is most commonly his situation in these islands, and indeed in the whole of Western Europe, for the centres of the majority of the depressions which visit us pass over the northern part of Great Britain, or else entirely clear of the north of Scotland.

When the observer finds himself on the northern side of the track he will experience a different sequence of phenomena. The wind will 'back' or shift against the sun, from South-east through East to North-east and North; and if he is near the centre the rain will fall steadily and heavily, and will not cease suddenly as soon as the barometer begins to rise.

As to an anti-cyclone, it will be perceived, from what has already been said, that no great changes of weather take place during its prevalence, except the occasional formation and clearing away of fogs.

The weather we experience in Western Europe is distinctly related to these areas of depression and anti-cyclones, to the rate at which they respectively travel over the earth's surface, and to the distance which intervenes between their respective centres. As in a system of either kind we may meet with winds from any point of the compass, which will have different qualities as to temperature, humidity, &c., according as they belong to one or the other, we see the great importance of the consideration, first pointed out by W. Köppen,[1] and subsequently by Capt. Toynbee,[2] that *the climatic character of a wind depends on its origin*, i.e. on its belonging to a depression or to an anti-cyclone. Weather windroses giving figures calculated without any regard to this fact are, therefore, unsatisfactory.

Again, the principle which Dove laid down under the name of the Law of Gyration (*Drehungsgesetz*), to the effect that the wind generally shifted with the sun; that a South-east wind shifted to West through South, and that any shift in the opposite direction was a sign of unsettled weather, simply means that stations in the

[1] *Repertorium für Meteorologie*, vol. iv., 1875.
[2] *The Meteorology of the North Atlantic during August*, 1873. London, 1878, p. 97.

west of Europe lie on the southern side of the paths of depressions, where the wind veers, or shifts with the sun. In Germany especially it is comparatively rare for a depression to pass on the southern side of a station, and this is one of the conditions which produce a 'backing,' or a shift against the sun, of the wind. If we examine the changes of wind in Iceland, we find backing is more common than veering, for there the depressions frequently pass to the south of the island. In fact, the evidence which Dove cited in support of his law is conclusive as to its truth in Europe, but not in other parts of the world.

Anti-cyclones are generally more or less stationary; but depressions move over the earth's surface, usually from west to east in these latitudes, their paths, as they advance, though chiefly ruled by the distribution of pressure, being liable to modification by the irregularities of the surface over which they pass; and their effects, as to the amount of cloud and rain to which they give rise, being influenced by the same causes. A South-west wind, for instance, may blow over a flat country with a clear sky, but as soon as the air reaches a hill side and is forced to ascend, the moisture it contains is condensed, clouds are formed, and rain is frequently the result.

I cannot here treat at any length of the subject of forecasting of weather. This, for a region situated like the United Kingdom, is a matter of great uncertainty, from the rapidity with which the changes succeed each other, and the extent to which local conditions affect the phenomena actually experienced.

The possibility of forecasting at all depends on the promptitude with which the earliest signs of a change of weather which show themselves at our western stations are telegraphed to the Central Office. Situated as the British Isles are, with an ocean to the westward, we, in London, can never get many hours' notice of a change. Our neighbours in France and Germany are better off, as they can usually receive intelligence of changes from the westernmost stations in the British Isles. Norway and Spain are nearly as unfavourably circumstanced as ourselves. In the United States the conditions are more favourable than in any part of Europe. There the Central Office is situated on the eastern side of a vast continent, and consequently reports can be collected from an extensive area, and the changes, as they come on, can be watched at head-quarters to an extent which is quite impossible at this side of the Atlantic.

What I have said relates more particularly to the temperate zone and to that part of it in which the British Isles are situated, but it is more or less true for all parts of the world. The region of which I have spoken is more frequently visited by depressions than other districts, and the changeability of its weather is proverbial. In countries where pressure is more steady the weather is proportionately more stable, and weeks may pass over without a change.

In the Trade-wind zones but slight changes of weather are experienced as long as the Trade blows; the air is comparatively dry, but when that current reaches a coast, especially if it be lined with mountains, heavy and constant rains are produced.

In Hindostan we have the alternation of the Monsoons; the North-east Monsoon, a dry wind, bringing with it persistent clear weather; and the South-west Monsoon, a wind which blows from the equator and comes over the sea, bringing deluges of rain to the Western Ghauts and the whole Coromandel coast, as well as to the flanks of the Himalayas, especially in the hill country of Assam.

In the Dutch East Indies we have similar conditions, excepting that, as the hemisphere is changed, the dry Monsoon is South-east, and the wet North-west.

FitzRoy, in his 'Weather Book,'[1] devotes four chapters to a description of the general features of weather over the globe, and since many parts of the world were practically known to him as a voyager of great and varied experience, and as a colonial governor, his summary will well repay perusal.

Before leaving the subject of weather, allusion must be made to the views which are held by many meteorologists, as to the existence of a periodicity in the changes of weather, related more or less closely to that which appears to be traceable in the sun's condition, as indicated by changes in the character of his surface. A brief summary of these views will be found in Appendix V.

[1] Longmans & Co., 1863.

CHAPTER XIX.

STORMS.

IN this, the concluding chapter of the book, I shall briefly discuss the principal features of the atmospherical disturbances which are known as Storms.

The forms in which these manifest themselves are very various. In some regions, and at certain seasons, the only breaks in the dull monotony of a tropical calm are sudden squalls of great violence but of short duration. Other regions resemble Kerguelen Island, where there is hardly an interruption to the persistence of the gales which sweep the Southern Ocean. In some parts, again, the wind, when it blows from a certain point, is liable to increase to the force of a storm, while in others the only type of storm which ever occurs is a revolving eddy, in which the changes of direction of the wind are the more rapid and extensive the nearer the station lies to the centre of the eddy.

Lastly we have the districts with which we are most familiar, situated in the higher latitudes of both hemispheres, where the storms are irregular in the frequency of their occurrence and in their character, occasionally blowing for a day or more without change of direction, while at other times the shifts of wind are almost as marked and sudden as in any of the eddies to which allusion has been made above.

The eddy type of storm is that which has attracted most notice, from the extreme violence which it manifests in certain regions, and the 'Law of Storms,' which Redfield, Reid, and Piddington discovered and popularised, is to be understood as applying to storms of this type.

The 'Law of Storms' is thus stated by Piddington:[1]—'The wind in hurricanes, and frequently in severe storms, in the higher latitudes on both sides of the equator, has two motions. It turns or blows round a focus or centre in a more or less circular form, and at the same time has a straight or curved motion forward, so that, like a great whirlwind, it is both turning round and, as it were, rolling forward at the same time. Next it is proved that it turns, when it occurs on the north side of the equator, *contrary* to the hands of a watch; and in the southern hemisphere that its motion is the other way, or *with* the hands of a watch.'

These storms are known under various local names. In the West Indies and the Southern States of the Union they are called Hurricanes. In the China Seas they are termed Typhoons. Piddington proposed for them the general name of Cyclones, owing to the supposed circular motion of the wind: a supposition which modern meteorologists have been obliged to abandon as not perfectly accurate, for the motion of the wind, in storms of the eddy type, is probably more truly spiral, incurving towards the centre, than circular.
On this point Piddington says:[2]—

[1] *Sailor's Hornbook*, 3rd edition, p. 8.
[2] *Ibid.*, p. 108.

'I should mention here that, though for convenience sake I have spoken of the wind as blowing in a circle, and the foregoing tables are calculated on that supposition, yet it is by no means certain that it is a *true* circle, or that even *if* the whole body of the storm be circular, the winds within it blow everywhere in exactly concentric circles.

'Mr. Redfield on this subject says, in a recent memoir ("Amer. Journ. of Science and Arts," 2nd Series, No. I., p. 14):—

'"When, in 1830, I first attempted to establish by direct evidence the rotative character of gales or tempests, I had only to encounter the then prevailing idea of a general rectilinear movement in these winds. Hence I have deemed it sufficient to describe the rotation in general terms, not doubting that on different sides of a rotatory storm, as in common rains or sluggish storms, might be found any course of wind, from the rotative to the rectilinear, together with varying conditions as regards clouds and rain.

'"But I have never been able to conceive that the wind in violent storms moves only in circles. On the contrary, a vortical movement, approaching to that which may be seen in all lesser vortices, aërial or aqueous, appears to be an essential element of their violent and long-continued action, of their increased energy towards the centre or axis, and of the accompanying rain. In conformity with this view, the storm figure on my chart of the storm of 1830 was directed to be engraved in spiral or involute lines, but this point was yielded for the convenience of the engraver."'

The localities where these storms appear developed in their greatest violence are well marked. In the first instance there is the West Indies, especially the line of the Windward Islands. Then come the corresponding latitudes of the western part of the South Indian Ocean, where the storms in their passage usually approach the island of Mauritius, and are called 'Mauritius Hurricanes.' The China Seas and the Bay of Bengal are also very liable to such visitations, and in the former district, as already stated, the storms are called 'typhoons.'

However, few parts of the sea in low latitudes are entirely free from cyclonic storms, and the following notice of the localities which are exempt from their visits, and of the general tracks which they follow wherever they occur, will be of interest:—

The only seas where they have as yet not been observed are the immediate vicinity of the equator in all oceans; and again, the South Atlantic Ocean as far as 25° S., and the eastern part of the South Pacific; in other words, the domain of the regular South-east Trade wind.

With these exceptions, these storms are met with in all seas between the parallels of 10° and 20° N. and S. latitude. In the northern hemisphere they travel in a general north-westerly direction, and in the latitude of 20° to 35° they turn at a sharp angle and advance to the north-eastward. The foregoing description applies to the southern hemisphere exactly, if south be substituted for north throughout. The track of the storm after recurvature cannot, however, be inferred from its direction within the Trade-wind zone, before it takes the turn.

In confined waters, such as the Caribbean Sea, the Bay of Bengal, and the China Sea, the paths are not so regular, but they follow the general directions indicated above.

The seasons at which the storms occur most frequently in the open sea are the summer and autumn of both hemispheres, as is very decidedly shown by the table given below. On the coasts of Hindostan they are commonest at the changes of the Monsoons in May and October.

Table of recorded Hurricanes, Typhoons, &c., in various parts of the World.

	Jan.	Feb.	March	April	May	June	July	August	Sept.	Oct.	Nov.	Dec.	Total
West Indies (300 years) (a)	5	7	11	6	5	10	42	96	80	69	17	7	355
South Indian Ocean (39 years, 1809–48) (b)	9	13	10	8	4	—	-		1	1	4	3	53
Bombay (25 years) (c)	1	1	1	5	9	2	4	5	8	12	9	5	62
Bay of Bengal (139 years) (d)	2		2	9	21	10	3	4	6	31	18	9	115
China Seas (85 years) (e)	5	1	5	5	11	10	22	40	58	35	16	6	214

The respective authorities are for (a) the Royal Geographical Society; for (b) Reid, Thom, and Piddington; for (c) (?); for (d) Blanford; for (e) Captain A. Schück.

The storm systems bear a strict resemblance to each other, the essential difference between the two hemispheres being that already explained, that in the northern the rotation of the wind is against watch-hands,

while in the southern it is with them,

In every case there is great barometrical disturbance, the barometer at the centre of the storm standing several tenths of an inch, sometimes nearly two inches, lower than outside the storm area. It is said that in the hurricane of Guadaloupe, September 6, 1865, the barometer at Marie Galante fell 1·693 inches (from

29·646 inches to 27·953 inches) between 6 h. 30 m. and 7 h. 40 m. A.M., *i.e.* in an hour and ten minutes. ('Buchan's Handy Book,' p. 266.)

No positive rule can, however, be given for the amount of barometrical depression to be expected in one of these storms, although Meldrum says that in the cyclones of the South Indian Ocean the barometer always falls below 28 inches.

No rule is, therefore, possible by which a man can determine his distance from the centre of the storm by means of readings of his own barometer alone; indeed, within the area of the storm itself it has on several occasions been found that smaller eddies existed, producing local and temporary barometrical minima, and the possibility of the existence of such disturbing causes makes it still more difficult for a captain who can have no means of knowing what changes in pressure are going on about him to draw conclusions from the readings of his own barometer. It is, however, a matter of the greatest importance to a captain to know when he is actually in danger of experiencing the full fury of one of these storms; and on this head Dr. Meldrum gives some very useful hints. The storms which he has had to deal with occur in the South-east Trade wind, which at times blows very strongly. How then is a captain to know when an increase of wind from the South-east, which he experiences, betokens the approach of a cyclonic storm and when not? Moreover, some of these disturbances are of less violence than others, and these may be weathered without fear of material damage. Again, if a captain finds himself at a distance from the path of the centre he may with great advantage avail himself of the strong winds on the external

B B

portions, without any risk of being drawn into the vortex.

Dr. Meldrum says, 'The most dangerous case of all is when the wind is steady from South-east, the barometer falling, and the wind gradually increasing in force. In this case, if the wind does not veer, and the barometer has fallen from the commencement fully sixtenths of an inch, I should advise him, as a last resort, to run to the north-westward, if possible.'

As to the handling of ships in these storms, it is evident from the constitution of the storm system that the wind in front must be directed across the path of the centre, blowing towards it on one side and away from it on the other. In fact, if we suppose the cyclone to be bisected by a line representing its track, and that another diameter be drawn perpendicular to this, the front quadrant (*d*, fig. 43), in which the wind blows towards the path, is the most dangerous position of all.

The French writers on these storms term the two quadrants *d* and *b le demicercle dangereux*, and the other two quadrants *a* and *c le demicercle maniable*. These two semicircles change sides when the hemisphere is changed, for the dangerous semicircle is always on the right-hand side of the path in the northern, and on the left-hand side of the path in the southern hemisphere, as will appear from the following diagram (fig. 44), borrowed from Rosser's 'Law of Storms.' In this the two halves are designated by the letters, D for 'dangerous,' and N for 'navigable,' the best translation of *maniable*.

It will be seen that in both cases the dangerous half is that on the side towards which the curvature of path takes place.

STORMS. 371

In the former of these semicircles, a ship running before the wind will be gradually brought nearer to the path and in front of the centre, and therefore into greater danger; so that a captain, in the front quadrant of the semicircle D, has to decide whether he

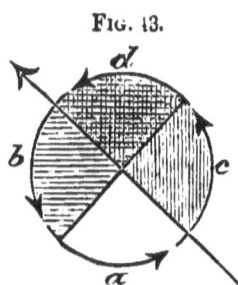

I. Northern Hemisphere.

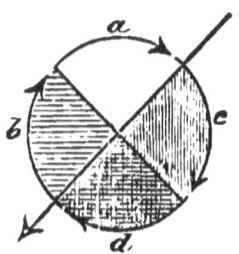

II. Southern Hemisphere.

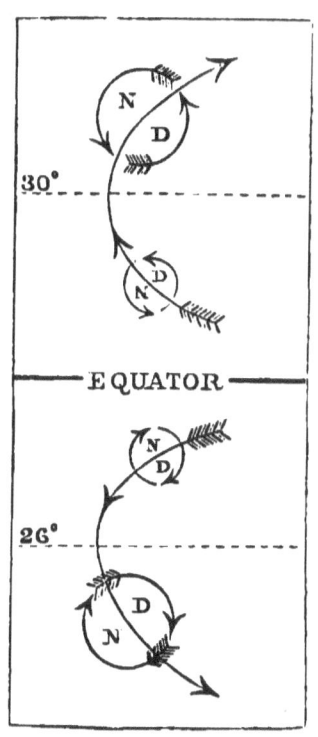

Fig. 44.

Tracks of Cyclones north and south of the Line.

will heave-to and let the storm pass him, or run before the wind so as to cross the path before the centre comes up to him. If he adopts the former course, he should lie-to on the starboard tack[1] in the

[1] For the information of the non-nautical reader, I may explain that 'on the starboard tack' means having the wind on the starboard side of the ship, i.e. on the right-hand side, looking towards the bow.

northern, and on the port tack in the southern hemisphere. Throughout the whole of the dangerous semicircle the wind will veer at a fixed station, which a ship lying-to practically is, as the storm passes on; throughout the navigable semicircle it will back. The tack suggested respectively for the semicircle D will in the two hemispheres lay the ship with her head away from the centre of the storm, will make her 'come up' and stem the sea, and enable her to avoid the risk of being taken aback as the wind shifts.

In the other half of the storm a ship running before the wind will be brought into greater safety.

Rules have been laid down by the Admiralty and others for the handling of ships in these storms under all the circumstances which can occur.

In all cases the most dangerous winds are those which are nearly at right angles to the path along which the storm is advancing. In the West India Hurricanes and the China Sea Typhoons this wind is North-east, within the Trade-wind zone, as the storms advance towards the north-west. In the South Indian Ocean it is South-east within the Trade-wind zone, as there the storms move towards the south-west. After the recurvature of the storms in the northern hemisphere they advance towards the north-east, and the most dangerous wind is the South-east. In the southern hemisphere a corresponding change takes place; they advance towards the south-east, and the most dangerous wind is the North-east.

The most important general facts about all these storms are that the wind blows in spiral curves, over most of the storm area, and that it increases in force as the centre of the system is approached. At the very centre there is a calm area, varying in diameter from about 100 miles to a very few miles, and even to yards, in the case of small whirlwinds.

On each side of this calm region the winds are blowing with great violence, and from directions nearly exactly opposite to each other. In nearly all accounts of cyclones it is reported that after the passage of the central calm the wind recommenced blowing with undiminished fury from the point exactly opposite to that from which it had blown before.

If the ship be on the path of the centre, the barometer falls continually and rapidly as long as the wind remains unchanged in direction; it is at its lowest when the ship is in the central calm, and when the wind comes on again from the opposite quarter the mercury rises again as fast as it had previously fallen.

This is the normal sequence of phenomena in an ideal hurricane, but it may fairly be said that no storm that has yet been investigated exhibits such regularity as has been indicated above. In the first place, from the very nature of the case, it is practically impossible to obtain observations from stations, or ships, in positions reasonably uniformly distributed over the storm area, so as to show that the wind was really blowing in uniform curves and with equal force all round the centre. The observations of a few ships have been taken, and *a circular form having been assumed for the storm*, the system has been laid down on the chart so as to show the observed winds as tangents to the isobaric curves.

The investigations of Meldrum for the South Indian Ocean have shown that in parts of certain storms the wind has blown directly towards the centre, and

Captain Schück, in his recent work,[1] has quoted twenty-five instances from various parts of the world where one ship was at the centre of a storm and other ships within its immediate influence, in order to test the question of indraft of the wind. He admits that the material which he has collected is insufficient both in quantity or quality to answer the questions: (1) What angle does the wind direction make with the isobar at the place? and (2) is this angle uniform in all azimuths?

The most complete investigation into this subject of indraft in a cyclonic storm outside the limits of the Trade-wind zone has been that carried out by Captain Toynbee, in the case of the storm which took place in the Atlantic August 24, 1873, near Bermuda. Captain Toynbee has shown that at three successive epochs in the same storm, between 10 P.M. on the 24th and 1 P.M. on the 25th, the general average angle of indraft from 108 observations was 29°, and the range of this mean indraft at the several epochs was from 25° to 31°. The ships were situated at varying distances, up to 700 miles, from the centre.

The practical result at which the most modern writers on the subject arrive is that it is not advisable to attempt any great precision in giving rules for the handling of ships in these storms. All that can be done is to give the captain rough and general rules for his guidance. It is certainly incorrect to assert that the centre of a cyclonic storm lies *exactly* at right angles to the direction of the wind. In some cases

[1] A. Schück, *Die Wirbelstürme, oder Cyclonen mit Orkangewalt, nach dem jetzigen Standpunkt unserer Kenntniss derselben.* 'Whirlwinds or Cyclones of Hurricane force, according to the present state of our knowledge thereof.' Oldenburg, Schulze'sche Buchhandlung, 1881.

cited by Meldrum the wind in parts of the storm area blew directly towards the probable position of the centre instead of blowing at right angles to the radius.

As has already been said, observations fail us if we wish to ascertain the actual phenomena occurring at the height of one of the West India hurricanes. At such a time houses are unroofed, and the general destruction of property and danger to life is so great that in several instances the observers have had to leave their posts, and self-recording anemometers, for registering the direction and force of the wind, have been damaged by the fury of the storm.

We are therefore unable to say whether the tropical cyclones exhibit the influence of different currents of air or possibly of a descending blast of cold and dry air forcing its way to restore atmospheric equilibrium within the rarefied region of the depression, which is the normal state of things at the close of one of our storms.

Accounts of the almost incredible violence of the wind in tropical cyclones will be found in many easily accessible works, and these need not, therefore, be reproduced here.

The storms are usually associated with torrents of rain and violent electrical manifestations, but of all their accompaniments the storm wave is beyond measure the most appalling in its effects. Where one of these storms advances against a low alluvial coast, the sea sometimes rises in a huge wave and sweeps over the country. This has happened on several occasions near Calcutta. In the cyclone of October 5, 1864, the water in the Hooghly Estuary rose 16·48 feet above the highest spring-tide level, covering the low-lying lands on the shores and occasioning great loss of life. This, however, fell far short of the destruction wrought by the Backergunge cyclone of October 31, 1876, at the eastern edge of the Ganges Delta. In this case the water covered the land to heights varying from 10 to 45 feet, as measured by marks on the trees, and the loss of human life exceeded, on the lowest estimate, 100,000 souls.

It has been already said that the West India hurricanes and other similar storms frequently pass out of the Trade-wind region, their tracks recurving to the eastward, when they enter the region of the prevailing Westerly winds, the Anti-trades. In such cases the area of the storm becomes more extensive and its action less destructive.

Some few of these storms have been traced from their point of recurvature the whole way across to the coasts of Europe, where they manifested themselves as ordinary storms. It is therefore evident that the ordinary storms of Europe are phenomena of the same nature as hurricanes, though differing from hurricanes not only in intensity, but probably in the shape of the storm area, and also apparently in the unequal development of the wind from the different points of the compass.

The storms of the temperate zones, many of which only differ in violence from tropical hurricanes, are very unequally developed on different sides; in these storms the area enclosed by the central isobar is very rarely circular, the opposite sides show strong contrasts between the temperature and hygrometric qualities of the winds found there, and lastly, one of their commonest features in the Northern Hemisphere is the sudden shift of wind from South-west or West to North-west,[1] p. 359,

[1] In the Southern Hemisphere the corresponding shift is of course from N.W. or W. to S.W.

simultaneously with a sudden upward movement of the barometer and downward movement of the thermometer.

The ordinary character of one of the storms in the west of Europe is that the central isobar is an irregular oval, and that while the gradients are steep and the winds proportionately strong on the southern side, the isobars are far apart and the gradients slight, on the northern, and the Easterly winds due to these gradients are consequently very light or else entirely absent.

The storms are unsymmetrical eddies, in which the wind always follows the Law of Storms and always keeps the lowest barometrical pressure on its left-hand side. The reason of the lack of symmetry is to be found in the distribution of mean barometrical readings, as shown in Plates VI. and VII. The normal condition of pressure in Western Europe in January is that the barometer ranges from about 30 inches over Northern France to 29·4 inches in Iceland. If, then, the centre of a depression, with a reading of 29·0 inches, lies near Holyhead—a very common case—there will be a gradient of 1 inch over about 400 miles from Holyhead to Brest; but one of only 0·6 inch over 800 miles from Holyhead to Reykiavik in Iceland. Accordingly, the East winds in Scotland can have but little force.

Some of these storms have an enormously elongated form. In the middle of February 1870, an Easterly gale of great violence prevailed over the southern part of the British Isles, and it was felt from the coast of Denmark to Corunna. Here was an apparent straight-line gale extending over twenty degrees of longitude, but it was really the wind of the northern side of a very elongated cyclonic depression.

All cases of so-called straight-line gales are to be

explained either by the persistence of the same characteristics for several days over the same region, or else simply by the fact that they are mere local phenomena, due to the contour of the country, like the exceptionally strong breezes often met with on rounding prominences and bluff headlands, or, to use a more familiar illustration, at street corners. By this statement it is not meant to imply that strong winds from a definite point are not met with for days together in the region of the Trade winds and Monsoons; but, firstly, the forces of these winds never reach those of a 'strong gale,' and, secondly, the conditions which cause them are of the same nature as those which cause our own storms, and it is the persistence of the conditions which determines the constancy of the wind in force and direction.

In some seasons a series of these storms passes in a long procession over the British Isles for weeks together, with usually two or three days' respite between the successive visitations. They move from west to east, speaking very generally, but under west we include the entire semicircle from south, round by west, to north. It is exceedingly rare in Western Europe to find a storm which exhibits a motion from any eastern point to any western.

As to the rate of the motion of storms, it varies between wide limits. At times depressions may be stationary for some hours, or even more than a day, while at times they may advance at the rate of fifty, sixty, or even seventy miles an hour during part of their course; this latter speed having been attained, *e.g.*, by the storm of March 12, 1876, in its passage over Northern Germany.

It is remarkable that the force of the wind has no traceable relation to the rate at which the system to which it belongs moves. The West India hurricanes, which are among the most violent of storms, travel very slowly, their average rate being about ten miles an hour. It is in our own latitudes that the maximum speed of translation has been observed.

It is evident that if we were to conceive of cyclonic depressions as simply revolving eddies advancing from west to east with a velocity v, and in which the air, at a distance of 100 miles from the centre, moved with a uniform speed s all round the centre, the velocity of the Westerly wind on the southern side of the system at the given distance would be $s + v$, and of the Easterly wind on its northern side $s - v$ at the same distance from the centre.

That any such relation exists is disproved by the fact that the Westerly wind blows quite as hard in stationary as in quick-moving depressions, and, *vice versâ*, we have had Easterly winds of great force in the latter case and an entire absence of them in the former.

The fact of the motion of storms renders Storm Warnings possible. When a storm shows itself on any part of the coast it is usually possible to issue, by telegraph, to the localities which it menaces, notice of its existence, and of the danger of its approach. As, however, meteorologists are at present unable to predict precisely the direction of motion or rate of advance of storms, there is necessarily considerable uncertainty about the warnings. Although for the British Isles it may be said that few storms reach the east coast before warnings have been issued, yet these are unfortunately the most violent and dangerous, owing to the extreme suddenness of their arrival. As to the west coast it must be admitted that the warnings for storms coming from the Atlantic are not unfrequently issued too late.

In all that has been said I have treated of storms as connected with simple systems of depression travelling over the earth's surface. These systems are, however, subject to constant modifications by processes of segmentation. Lesser eddies are found on the outskirts of the original depression, which modify it and are themselves modified by it. At times these latter 'secondary,' 'subsidiary,' or 'satellite' depressions, as they are called, develop greater energy than their primaries, often travelling faster, and being proportionately more dangerous and destructive. If we could be certain that a system of depression, once formed, would preserve its character and track unchanged for a number of days, the problem of issuing warnings would be comparatively simple; but this is very far from being the case, and the constant changes in the size, shape, track, and general characteristics of a storm system are among the greatest difficulties experienced in the way of correct forecasting of storms.

The motion of depressions is more or less dependent on the general distribution of pressure, but all that can be said here is that this motion appears to be regulated by the position of the areas of high pressure, the anti-cyclones. The depressions have a tendency to revolve round the anti-cyclones, at least, speaking of the north temperate zone, round their western, northern, and eastern sides; but, as already stated, they very rarely move from east to west in these latitudes, so that they do not pass for any distance along the southern edge of an anti-cyclone, although the hurricanes of low latitudes, as of the West Indies, for instance, do move to the Westward along the southern edge of an area of high pressure. Inspection of the maps of pressure distribution (Plates VI. and VII.) will show how the general motion of storms over the Atlantic is related to the permanent area of high pressure, about the fortieth parallel of latitude.

I cannot here enter at greater length into the subject of Storm Warnings or Weather Telegraphy, but these are more fully treated of in the work to which I have already referred, 'Weather Charts and Storm Warnings.'

Closely related to hurricanes and typhoons, and only differing from them in size, are the various forms of whirlwinds popularly, but erroneously, called Tornadoes, a name properly belonging to certain squalls on the west coast of Africa, which will be described subsequently.

Whirlwinds appear more frequently in continental countries, but are also occasionally experienced in the British Isles. Their diameter varies from a few feet to perhaps a mile or so, and the violence of the wind in them is often fully equal to that of a West India hurricane. One which passed over part of Kent, between Walmer and Deal, on October 24, 1878, left behind it a track varying from 450 to 700 feet in width and extending for more than a mile, over which everything had been destroyed; in some cases half of a house had been blown down, the other half remaining uninjured. Very commonly whirlwinds exert a strong lifting action, raising heavy objects, such as branches of trees, carts, &c., off the ground and carrying them to some distance. They are often of very limited vertical extension, as is shown by the instance of a whirlwind which passed over the village of Hallsberg in the province of Nerike, in Sweden, August 18, 1875. In this case it was noticed that while the branches of trees and fragments of the wreck of buildings were carried by the wind for miles, the clouds, even at a comparatively low altitude, did not indicate the slightest sign of disturbance.

Whirlwinds in passing over a dry plain often sweep up a quantity of dust, and the phenomenon assumes the shape of a vertical revolving pillar accompanied by great electrical disturbance. If the track crosses a sheet of water or even a river, a water-spout is formed during the transit, as at Königswinter, on the Rhine, June 10, 1858.[1]

The passage from whirlwinds to dust storms on land and to waterspouts at sea is easy. These are phenomena limited in extent and transient in their character, and they occur principally in tropical countries. The rotation of the air in them is sometimes with, sometimes against, watch-hands.

In various parts of the world gales, often of great violence, are known by local names, but their true character may be easily understood from what has been said in the preceding pages.

The most generally known of these disturbances, and the most serious in its effects, is the Simoom of the Arabian desert, which at times buries whole caravans

[1] Reye, *Die Wirbelstürme*, p. 30.

in sand, and of which graphic descriptions will be found in books of travel relating to the countries subject to its visitation. It usually only lasts for ten minutes or so.

The 'Nortes' of the Gulf of Mexico are Northerly winds, which blow with great force, and are often dangerous to shipping. The researches of American meteorologists have shown that these are only the winds on the western sides of areas of depression, which have not become developed into complete hurricanes.

The 'Mistral' of Provence, and the 'Bora' of Trieste, which have been already mentioned (p. 292), are also winds usually due to the presence of cyclonic disturbances, like the 'Nortes,' but they are aggravated by the fact of their coming down off high land; so that their origin is, at least in part, attributable to an action similar to that which produces the hill and valley winds, which have already been described (p. 288). Dr. Hann has recently pointed out that a violent Bora has been experienced at Trieste when the isobars alone had given no indication of danger.

On the Pacific side of Central America similar winds are experienced, and owe their origin to similar causes. They blow from North-east and North-north-east on the coasts of Nicaragua and Guatemala, and are known under the names of 'Papagayo' and 'Tehuantepec' winds, from the districts where they are respectively experienced.

The 'Pamperos' of Buenos Ayres are South-west winds, blowing over the Pampas of the Argentine States. They correspond to our North-west winds, and, like them, appear in the rear of depressions. They are often of extreme violence, dismasting vessels if sail has not been shortened on the first symptoms of their approach.

Lastly, there are the various forms of squalls, of the true nature of which much remains to be learnt, owing to the fact that they are of small extent and transient, so that we cannot easily obtain observations from various ships in different parts of the same squall.

The 'White Squall' is only so called from its whitening the sea with foam and spoon drift as it advances over the surface.

The true Tornado of the west coast of Africa, from about Cape Verde, along the Gold Coast, into the Bight of Benin, is a very remarkable squall, thus described by an officer who served for several years on the coast in a cruiser on the look-out for slavers:—

'Their approach is indicated by a well-defined arch of dark clouds, and so they give ample warning. They are of brief duration, and *generally blow off-shore*. The brush of wind out of the arch as it reaches a vessel, is of the most furious nature, accompanied by thunder, lightning, and the heaviest conceivable rain. I have a vivid recollection of at least half a dozen of these squalls, one, from the amount of thunder and lightning, the most awful spectacle in this line I ever witnessed.'

From the mention of the arch of clouds, it is evident that these gusts of wind are identical with what are known in other seas as 'arched squalls,' and which, like tornadoes, are accompanied by violent electrical phenomena and torrents of rain.

The barometer does not give much signs of their approach, though it generally falls, and indicates a reduction of pressure of about a tenth of an inch at the height of the squall. Captain Schück, in a paper on 'Arched Squalls,'[1] speaks of evidence of a decided shifting of wind in the squall. 'The African Pilot,' Part I. (3rd edit., 1880), at p. 300, says, 'The tornadoes generally commence from the S.E., and draw round through East to N.E. : but this is by no means certain, as they have been experienced blowing right on shore. They have been known to begin at East and draw round to South, and they have also been experienced from the Westward.'

This description applies to the eastern part of the district of their prevalence, in the Bight itself. In the western part, near Sierra Leone, the direction is *always* off-shore, and the officer who kindly furnished the description given above adduces very telling evidence of this. He says that the cruisers always lay close in-shore, because the captains knew that there was no danger at all of that shore turning into a lee shore, and that they might always run before the wind of the tornado with plenty of sea room and without any fear of stranding.

[1] *Annalen der Hydrographie*, March 1877, translated in the *Quarterly Journal of the Meteorological Society*, vol. v. p. 75.

APPENDICES.

Explanation of the Tables in Appendices I., III., and IV., extracted from 'Instructions in the Use of Meteorological Instruments.' London, 1875.

APPENDIX I. shows the value of readings of thermometers graduated on Fahrenheit's scale, in terms of the centigrade scale, and Reaumur's scale.

APPENDIX III. contains corrections to be applied to the readings of barometers mounted in *brass* frames, in order to reduce them to the normal temperature, 32°. They have been computed from the following formula given by Schumacher :—

$$\text{Correction} = -h \frac{m(t-32) - s(t-62)}{1 + m(t-32)}$$

in which
h = reading of the barometer,
t = temperature of attached thermometer,
m = expansion of mercury for 1°F., taken as ·0001001 of its length at 32°,
s = expansion of the substance of which the scale is made; for brass s is taken as ·00001041 of its length (h) at the standard temperature for the scale, viz., 62° F.

APPENDIX IV. is for reducing to the sea-level observations of the barometer made at any height not exceeding 1,000 feet. It is given for two pressures at the lower station, namely, 30 and 27 inches. For intermediate pressures and heights the correction may be obtained by proportional parts.

The value at the sea-level, of a barometer reading, at a station, the height of which is known, may be calculated from the following formula :—

$$\text{Log} \frac{h}{h'} = f \div \left\{ 60159 \left(1 + \frac{t+t'-64}{900}\right)\left(1 + \cdot 00268 \cos 2\, l\right) \left(1 + \frac{f+52251}{20886861}\right) \right\}$$

From a table of common logarithms, the natural number corresponding to $\log \frac{h}{h'}$ is found; or, $\frac{h}{h'} = n$,

And $h = n\, h'$.

In this formula—

h and h' = barometer reduced to 32° F., at the lower and upper stations respectively,

t and t' = the temperature of the air at the respective stations,

f = elevation of the upper station in feet,

l = latitude of the place.

The above formula is merely an inversion of the well-known formula given by Laplace in his 'Mécanique Céleste,' for finding the difference of elevation between any two places by means of the barometer, which, adapted to Fahrenheit's thermometer and English feet and inches, is,—

$$f = 60159 \log \frac{h}{h'} \left(1 + \frac{t+t'-64}{900}\right)\left(1 + \cdot 00268 \cos 2\, l\right) \left(1 + \frac{f+52251}{20886861} + \frac{x}{10443430}\right)$$

In this formula f is the difference of elevation between the two stations, and x is the height of the lower station above the sea-level.

In the last factor an approximate value must be used for f.

APPENDIX I.

COMPARISON OF THERMOMETER SCALES.

Fahrenheit	Centigrade	Reaumur
+ 212.°	+ 100.°0	+ 80.°0
100·	37·8	30·2
90·	32·2	25·8
80·	26·7	21·3
70·	21·1	16·9
60·	15·6	12·5
50·	10·0	8·0
40·	+ 4·4	+ 3·5
32·	0·0	0·0
30·	− 1·1	− 0·9
20·	− 6·7	− 5·3
+ 10·	− 12·2	− 9·8
0·	− 17·8	−14·2
− 10·	− 23·3	−18·6
− 20·	− 28·9	−23·1
− 30·	− 34·4	−27·5
− 40·	− 40·0	−32·0
− 50·	− 45·5	−36·4
− 60·	− 51·1	−40·9
− 70·	− 56·6	−45·3

Value of one degree of the scales:—

1° Fahrenheit = $\frac{5}{9}$ Centigrade = $\frac{4}{9}$ Reaumur
1° Centigrade = 1·8 Fahrenheit = 0·8 Reaumur
1° Reaumur = 2·25 Fahrenheit = 1·25 Centigrade.

APPENDIX II.

MODE OF CONSTRUCTION OF THE CHRONOISOTHERMAL DIAGRAM, P. 48.

The curves are drawn for every 2° Fahrenheit, and the monthly means are referred to the centres of the month spaces.

A square of the required size is constructed, and divided vertically into twelve divisions, for the months of the year, and horizontally into twenty-four divisions, for the hours of the day.

Above this square a working diagram is drawn, having similar divisions for the months, but being divided horizontally into degrees instead of hours.

The mean temperature of any given hour for the several months is then entered on this diagram by dots, in the centres of the monthly spaces, at the required heights, as shown by the side scale of degrees. These dots are then joined by lines so as to form a continuous curve, representing the mean temperature at the given hour throughout the year.

The points where this curve cuts the *even* degree lines are projected downwards on the line in the square which corresponds to the hour for which it is drawn.

The same process is repeated for each of the twenty-four hours.

A second working diagram is then drawn at the side of the square, divided horizontally into hours corresponding to the hour lines in the square, and vertically into degrees.

Upon this scale curves are drawn representing for each month the hourly range of temperature, and the dots are projected laterally on the line in the square which corresponds to the month.

The points in the square which represent the same degree of temperature are then joined, and the result is a diagram like that shown in fig. 8.

APPENDIX III.

Correction to be applied to Barometers with *Brass Scales* extending from the cistern to the top of the mercurial column, to reduce the observations to 32° Fahrenheit.

Temp.	27 inches	30 inches
0°	+ ·069	+ ·077
28	+ ·001	+ ·001
29	− ·001	− ·001
30	− ·004	− ·004
40	− ·028	− ·031
50	− ·052	− ·058
60	− ·076	− ·085
70	− ·100	− ·111
80	− ·124	− ·138
90	− ·148	− ·164
100	− ·172	− ·191

APPENDIX IV.

Table for reducing Observations of the Barometer to Sea-level.

Correction additive.

Height in Feet	Temperature of External Air		
	0°	40°	80°
I. When the Barometer reading at Sea Level is 30 inches.			
10	+ ·012	+ ·011	+ ·010
100	·123	·112	·103
1000	1·208	1·105	1·017
II. When the Barometer reading at Sea Level is 27 inches.			
10	+ ·011	+ ·010	+ ·009
100	·111	·101	·093
1000	1·087	·994	·915

APPENDIX V.

NOTE ON THE RELATION BETWEEN SUNSPOTS AND WEATHER.

When the sun is observed through a telescope, its disc is usually found to exhibit spots on different parts of the surface. These spots are not fixed objects, like the mountains seen in the moon, but are variable, not only in number, but in position, configuration, and dimensions.

Sometimes these spots are numerous and large, covering, for a period, millions of square miles of area; at other times they are few in number and insignificant in size. Occasionally days may pass without any being perceptible through the most powerful telescopes.

Although these spots were discovered almost at the time of the invention of the telescope, they only became objects of *systematic* study in 1825, when Hofrath Schwabe, of Dessau, commenced to observe and record them daily; a task he continued to perform, almost without intermission, till 1867.

This observer found that the epochs at which spots were most frequent and largest were separated by an interval of about ten years, and that an interval of nearly equal duration separated the epochs of greatest rarity of spots. More recent determinations indicate that the length of the period is more nearly eleven years; but the data at present available are insufficient to establish satisfactorily its precise duration.

In 1852 Sir E. Sabine, in discussing magnetic observations made in different parts of the globe, found that a similar periodicity existed in certain of the phenomena of terrestrial magnetism; and modern observations show that the appearance of an exceptionally large spot on the sun's surface is almost invariably accompanied by a 'magnetic storm,' felt simultaneously in all parts of the globe.

When such magnetic storms occur, brilliant displays of the

Aurora usually take place; and hence it has been found that the appearance of the Aurora exhibits a periodicity allied to that of sunspots.

Inasmuch as such facts as have been mentioned point to a close connection between solar and terrestrial phenomena, physicists have been led to inquire whether meteorological phenomena may not be affected by the same periodicity.

Assuming that the amount of solar heat reaching the earth must be influenced by the extent of the sun's surface darkened by the spots, or else by the augmented solar energy indicated by the development of a great number of spots, it has been conceived that the forces concerned in the production of meteorological changes should be similarly affected, and that meteorological phenomena should exhibit an eleven-year period in the frequency of their occurrence.

It is maintained by many investigators of high authority—such as, among Englishmen, Meldrum and Balfour Stewart, not to mention foreigners—that such a periodicity does exist in the occurrence of cyclones, rainfall, the conditions of terrestrial temperature and of barometrical pressure, as well as in that of events depending on the foregoing—such as floods, famines, good or bad harvests and vintages, the price of grain, and commercial panics.

Inasmuch, however, as it is admitted that the connection between these phenomena and those of sunspot frequency is not sufficiently understood to justify prediction; and inasmuch as various investigators arrive at contradictory conclusions as to the nature of the connection: one class holding that sunspot frequency accompanies a high, while the other asserts that it is associated with a low temperature—it can scarcely be said that the close relation between solar and terrestrial phenomena is capable of accurate demonstration.

We may venture to surmise that, whenever the connection may be really discovered, it will prove to be of a less simple nature than has hitherto been supposed.

For the convenience of reference is subjoined a copy of the table compiled by Dr. Rudolf Wolf, of Zürich, indicating the relative extent of solar spots in each year. This table is generally employed as the basis of the calculations just referred to.

Wolf's numbers, indicating the relative values of the sunspots observed in the different years of the present century.

Year	Number	Year	Number	Year	Number
1800	15·3	1828	62·5	1855	6·7
1801	34·0	1829	67·3	1856	*4·3*
1802	55·0	1830	**70·7**	1857	22·8
1803	71·2	1831	47·8	1858	54·8
1804	**73·1**	1832	27·5	1859	93·8
1805	47·6	1833	*8·5*	1860	**95·7**
1806	28 9	1834	13·2	1861	77·2
1807	9 4	1835	56 9	1862	59·1
1808	7·7	1836	121·8	1863	44·0
1809	2·5	1837	**138·2**	1864	46·9
1810	0·0	1838	103 1	1865	30·5
1811	1·4	1839	85 8	1866	16·3
1812	5·5	1840	63·2	1867	*7·3*
1813	12·8	1841	36·8	1868	37·3
1814	14·4	1842	24·2	1869	73·9
1815	35·4	1843	*10·7*	1870	**139·1**
1816	**46·4**	1844	15·0	1871	111·2
1817	41·5	1845	40·1	1872	101·7
1818	30·0	1846	61 5	1873	66·3
1819	24·2	1847	98·4	1874	44·6
1820	15·0	1848	**124 3**	1875	17·1
1821	6·1	1849	95·9	1876	11·3
1822	4·0	1850	66 5	1877	12·3
1823	*1·8*	1851	64·5	1878	*3·4*
1824	8·6	1852	54·2	1879	6·0
1825	15·6	1853	39·0	1880	32·3
1826	36·0	1854	20·6	1881	54·2
1827	49·4				

NOTE.—The values for the years 1800 to 1865 are from the 'Astronomische Mittheilungen,' von Dr. Rudolf Wolf, No. L. for February 1880. Those from 1866 to 1881 from the 'Astronomische Mittheilungen,' No. LV. April 1882.

The maximum years are shown in Egyptian figures, thus (**73·1**). The minimum years in sloping figures, thus (*1·8*).

Plate I.

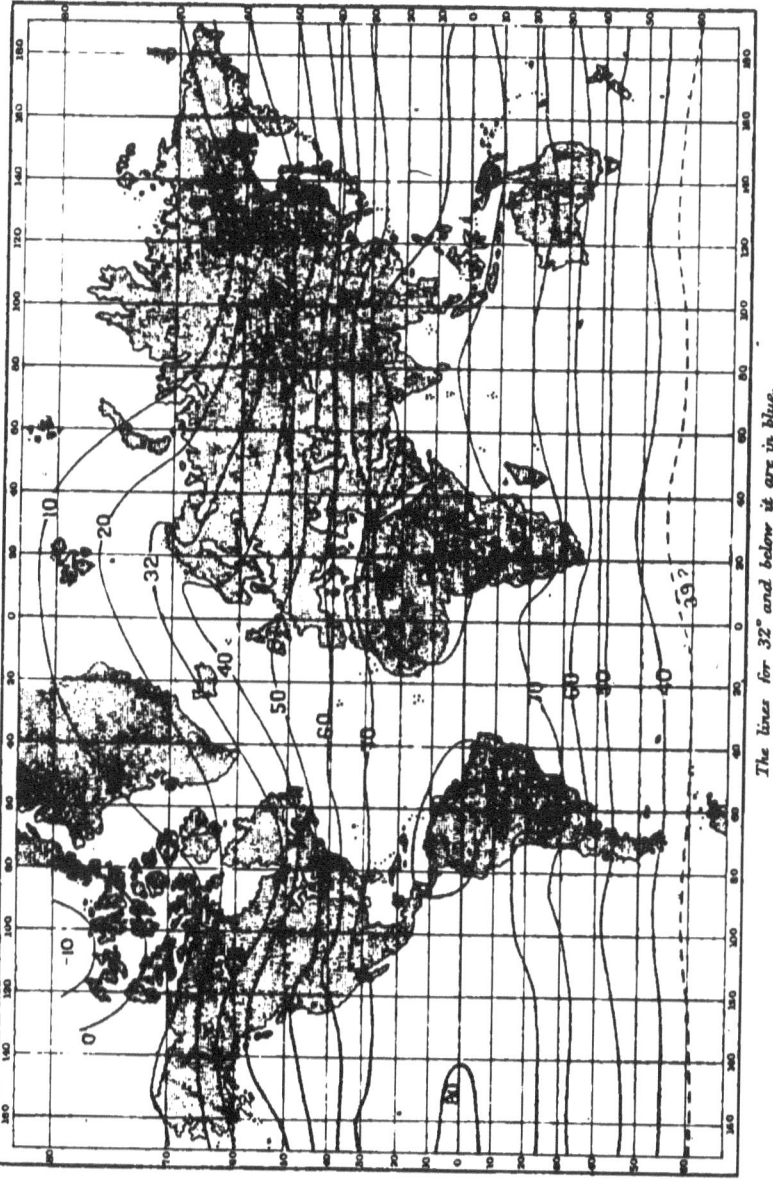

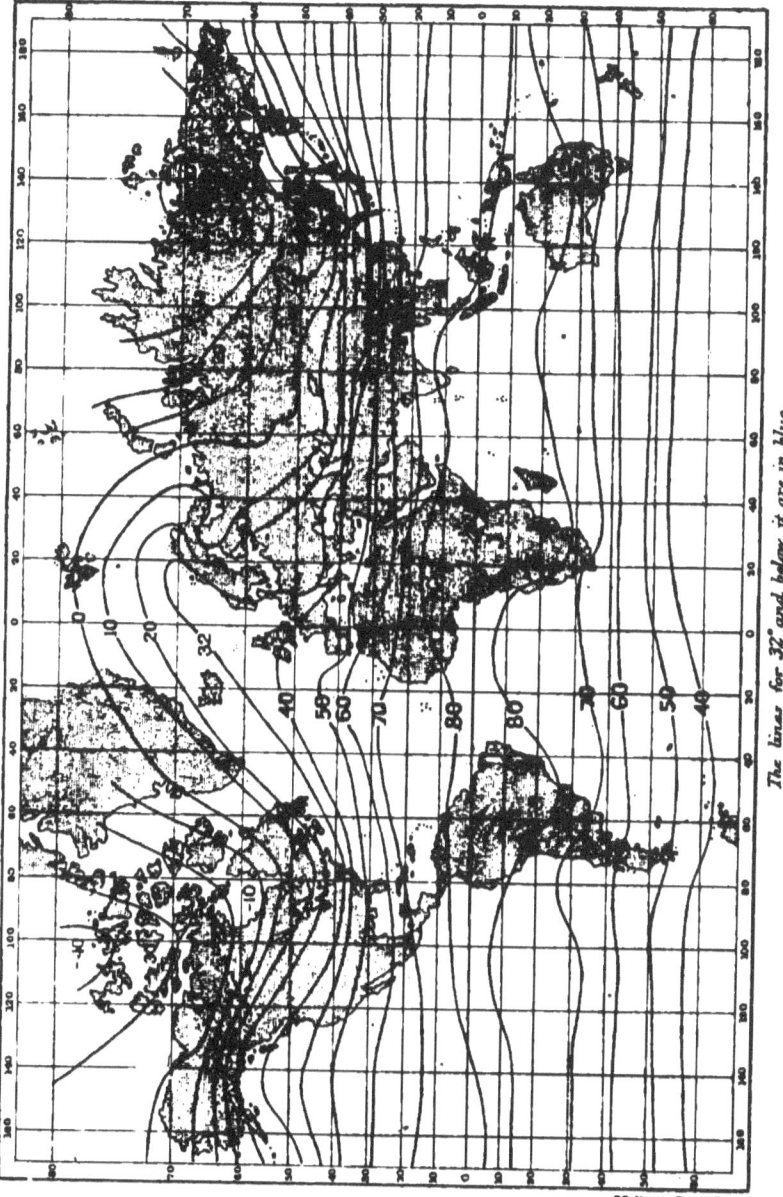

Plate. II.

The lines for 32° and below it are in blue.

Malby & Sons, Lith.

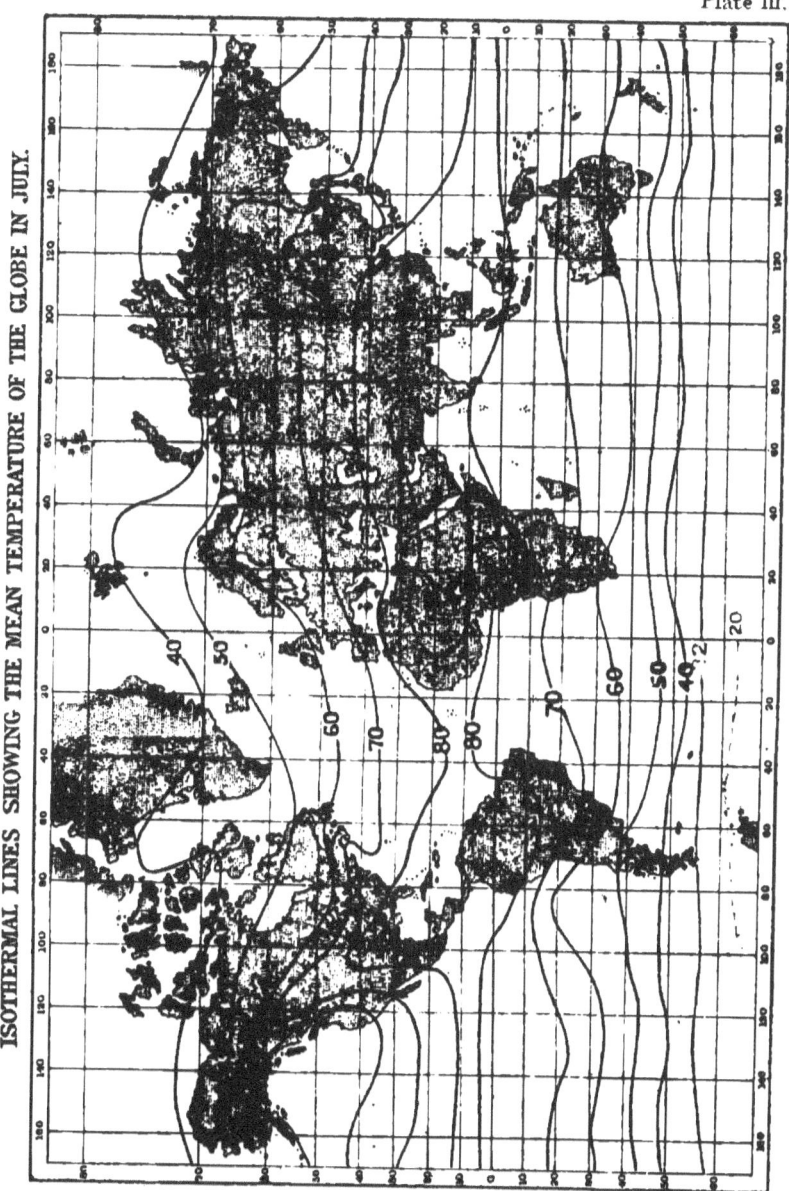

Plate III.

ISOTHERMAL LINES SHOWING THE MEAN TEMPERATURE OF THE GLOBE IN JULY.

Malby & Sons, Lith.

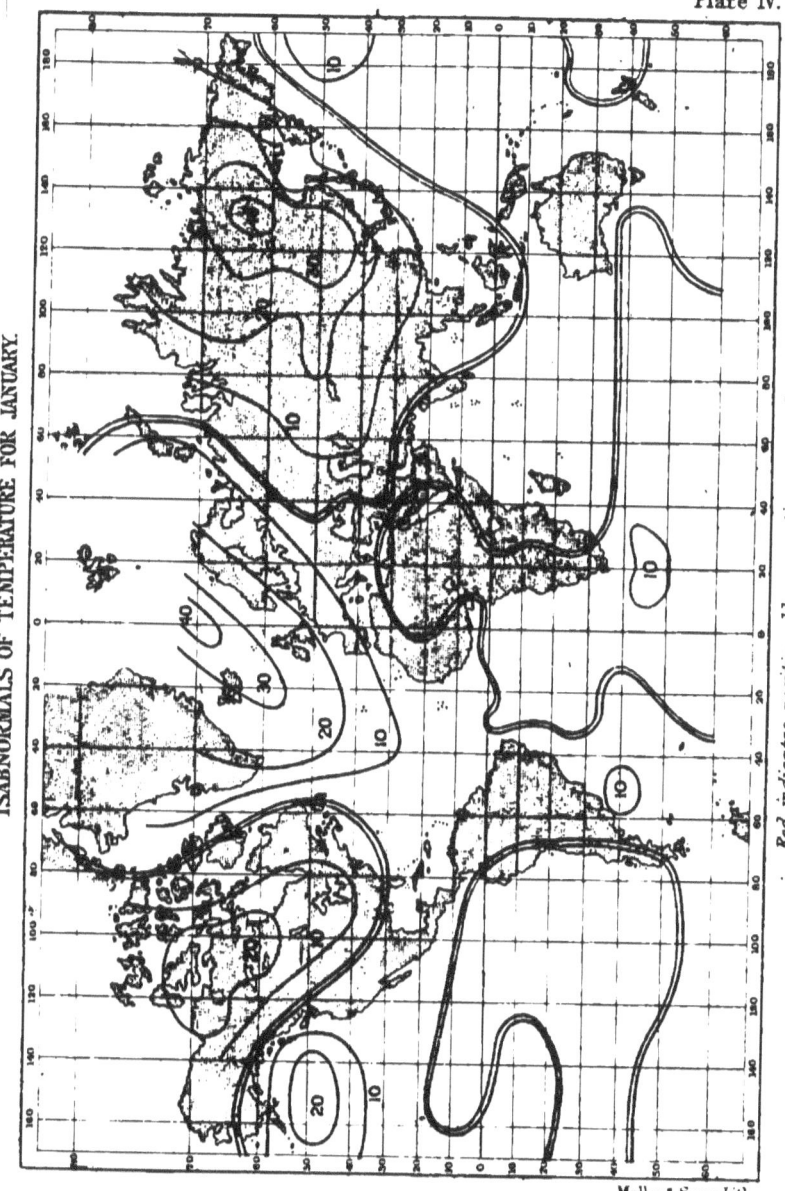

ISABNORMALS OF TEMPERATURE FOR JULY.

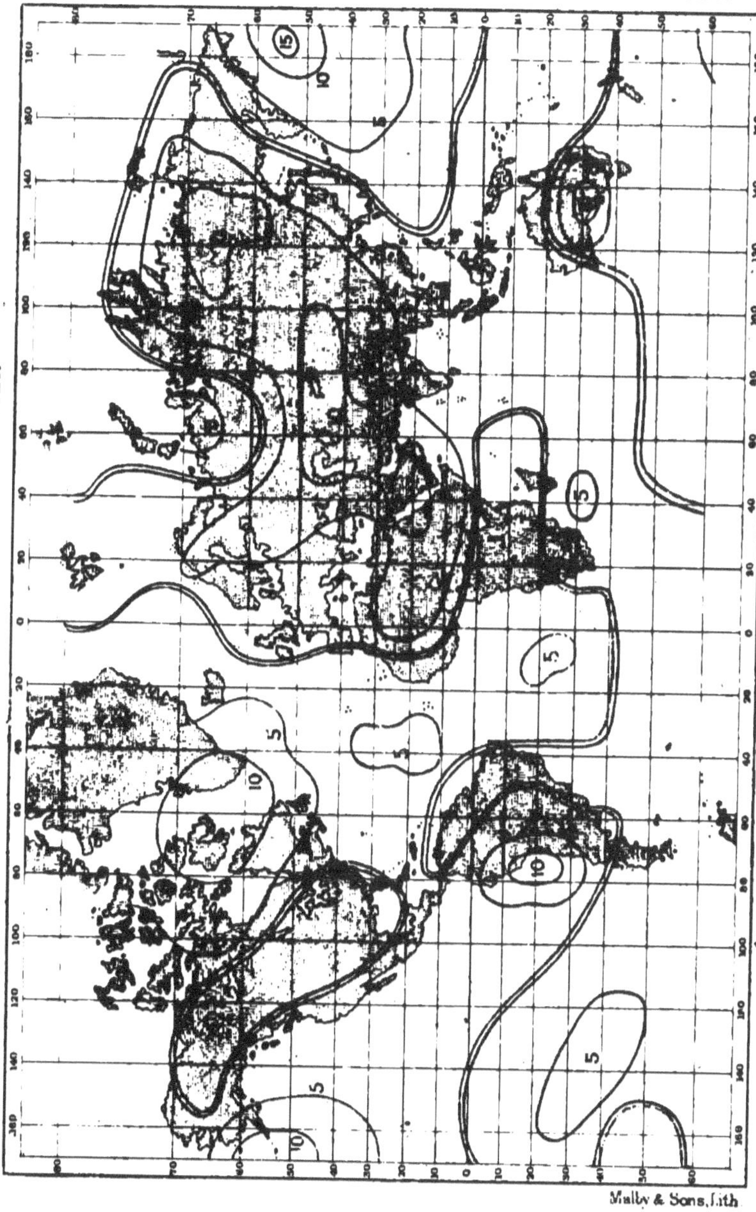

Plate V

Red indicates positive, blue negative, anomaly.

Malby & Sons, Lith.

Plate VI.

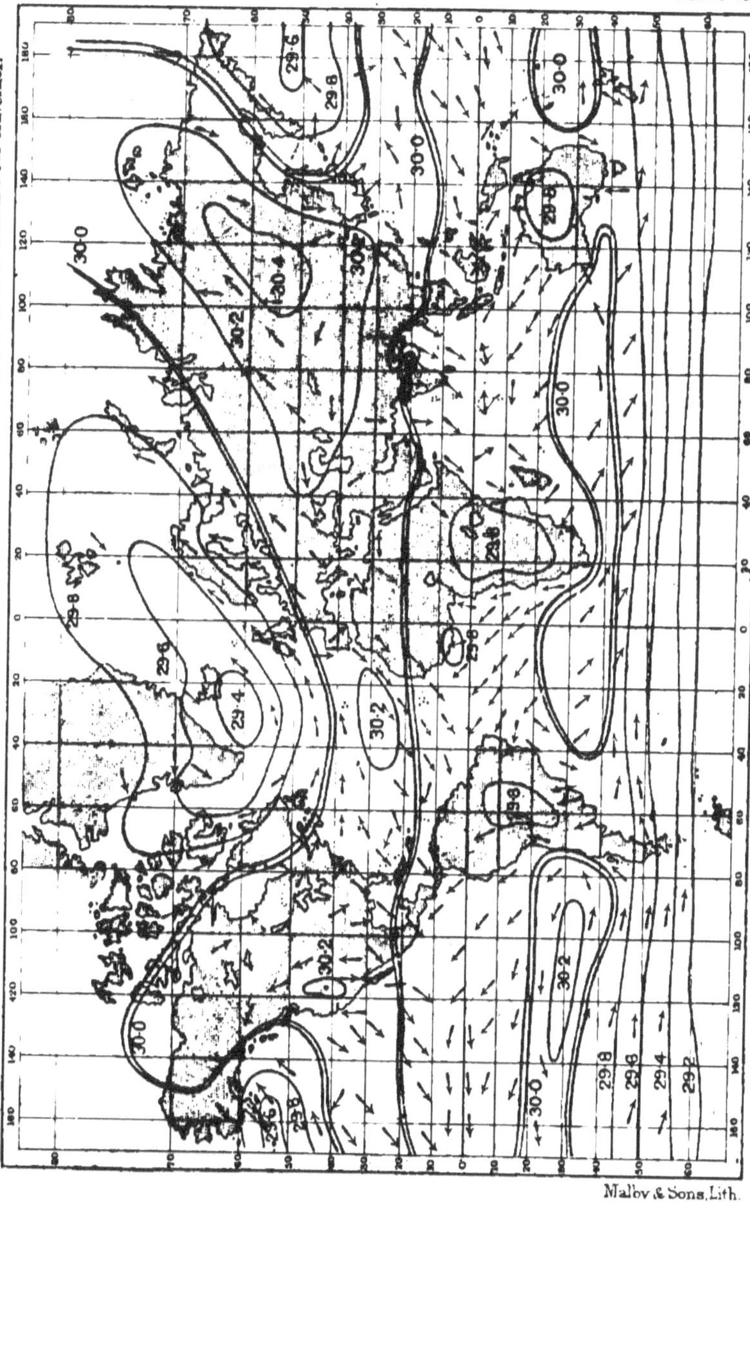

LINES SHOWING THE MEAN BAROMETRICAL PRESSURE & THE PREVAILING WINDS OF THE GLOBE FOR JANUARY.

Malby & Sons, Lith.

LINES SHOWING EQUAL ANNUAL RANGE OF TEMPERATURE FOR THE GLOBE.

INDEX.

ABI

ABICH, Prof., on hailstones, 143
Absolute humidity, 110
Absorption of sun's heat by atmosphere, 57, 211
Abyssinia, thunderstorms in, 191
Actinometer, Herschel's, 52; Balfour Stewart's, 52
Adams (W. G.), on return shock of lightning, 182
Admiralty, the, wind and current charts, 282, 314
'Advance' (Lieut. de Haven), drift of the, 309
Agulhas current, the, 308
Air, atmospheric, effects of change of level on, 288
— in tube of barometers, 82; detection of, 83
— motion in depressions and anti-cyclones, 357; in cyclones, 366, 369, 374
Aitken, on fogs, 120
Alaska, contrast of climates in, 306
Alnwick, lanes cut in the woods by a storm at, 155
Alps, observations of radiation in the, 57
Altitude correction for barometers, 85, 391
America, South, rainfall of, 328; climate of the south point of, 343
Andes, rainfall of the, 328
Andries, on causes of low pressure near South Pole, 251
Anemometer, Hooke's, 150; Wild's, 150; Lind's, 150; Osler's, 150;

ARM

Cator's, 151; Robinson's, 151; theory of Robinson's, 152; difficulty about satisfactory exposure of, 153
Aneroid, the, 79; defects of the, 80
Animals, distribution of, 339, 342
Annual range of temperature, 216, 339; of pressure, 93; of wind velocity in British Isles, 164; of thunderstorms, 191
Anomaly, the thermic, 232
Antarctic regions, low pressure over the, 251; ocean currents in the, 298, 309
Anthelia, 203
Anti-cyclones explained, 356; air motion in, 357; contrasted with depressions, 357; slight motion of, 361
Antisana, pressure at, in winter and summer, 241
Anti-trades explained, 245, 264; Laughton on, 274
Apjohn, formula for dry and wet bulb hygrometer, 108
Arabs, the, their knowledge of monsoons, 266
Arago, on globular lightning, 175
Arctic regions, solar radiation in, 57; thunderstorms in, 191; snow drift in, 327
— current, the American, 298; in Pacific, 306; in Polar regions, 309
Aristotle, on monsoons, 266
Armagh, annual period of wind velocity at, 164

ASC

Ascending currents, action of, in producing rain, 138, 288
Ascension, diurnal range of temperature at, 216; annual range at, 218; rainfall at, 321
Ashes from volcanoes borne by upper currents, 243
Asia, Central, annual range of pressure in, 93, 259
Astronomy, contrasted with meteorology, 1
Atlantic Ocean, motion of air over the, 352; limits of trade winds in, 246; currents of the, 299; rainfall of the, 327; the central, annual range of pressure over, 93
Atmometer, v. Lamont's, 100
Atmometry, Schmid on, 101
Atmosphere, height of, 12; constitution of, 13; variations in constitution, 15; circulation of, explained, 240; overflow of, near equator, 242. *See* Air.
Attached thermometer in barometers, 70
August, on hygrometry, 108
Aurora, description of, 191; noise of, 192; the 'corona,' 192; the dark segment, 192; colour of, 192, 193; at low levels, 193; height of, 193; geographical distribution of, 194; in southern hemisphere, 195; relation of, to weather, 195; relation of, to magnetism, 196
Autumn rains, regions of the, 332

BACKERGUNGE cyclone, 376
Backing of wind, the, 361
Baer, v. on contrast of climates on the coasts of Alaska, 306
Baikal Lake, effect of, on temperature, 225, 228, 229
Balloon ascents, Glaisher's, 12, 66
Barker, Lady, on winds of New Zealand, 290
Barnaoul, annual range of pressure at, 93; of temperature at, 218; diurnal range of temperature at, 216

BOO

Barograph, Brooke's, 77; Ronald's, 77; Redier's, 77; King's, 78
Barometer, invention of the, 63; Torricelli, 63; Pascal, 65; best material for scale, 66; graduation of the, 67; vernier, 67; mode of setting, 68; attached thermometer, 70; management of, 71; verification of, 74; defects of, 82; air in tube of, 82; boiling of, 82; pipette, 83; reasons for its rise or fall, 248
— corrections; capacity, 73, 84; capillarity, 84; index error, 84; temperature (reduction to 32°), 85, 391; altitude (reduction to sea level), 85, 391; diurnal range of, 88; Buchan on, 91; Eaton on, 90; annual range of, 93
Barometers, standard, 72; Fortin's, 72; Kew pattern, 73; marine, 74; siphon, 75; wheel, 76; aneroid, 79; metallic, 80
Barometrical pressure, on causes which produce changes in, 248; regions of reduced, 250; relation of, to wind, 253, 353; relation of to isabnormals of temperature, 257
Batavia, rainfall of, 323
Bearings of wind, true and compass, 148
Beaufort's wind scale, 158; velocities corresponding to, 159
Beckley, automatic rain gauge, 132
Behring's Strait, checks course of currents, 298, 318
Bidston Observatory, exceptional pressure of wind at, 156
Binnie, on evaporation at Nagpoor, 99
Black bulb thermometer *in vacuo*, 53; its use, 55
Blanford, on monsoons, 264; on the soft place in the monsoon, 268; on land and sea breezes, 286
Boiling of barometers, 82
Boiling point of water, 21; its use in hypsometry, 21
Boothia Felix, diurnal range of pressure in, 89; of temperature, 216

INDEX. 397

Bora, the, 292, 383
Bossekop, aurora at, 192
Bourdon's metallic barometer, 80
Bourke, Commander E. G., on cold currents of Atlantic, 303
Brazil, daily thunderstorms in, (Caldcleugh), 130, 147, 191; current, the, 305
Breezes, land and sea, 285; Blanford on, 286
Brewster, Sir D., on poles of cold, 234
British Isles, the, annual variation of wind velocity in, 164; distribution of winds in, 276; sea temperature on the coasts of, 319; weather telegraphy in, 362
Brooke's barograph, 77
Brussels, laws of atmospheric electricity at, 171
Buchan, A., on diurnal range of barometer, 91; on predicting hoar frost, 117; isothermal charts, 222; isobaric charts, 238
Buitenzorg, rainfall of, 323
Buran, the, 293
Buys Ballot's law, 255, 353

CALCUTTA, diurnal range of pressure at, 89; annual range of temperature at, 218
Caldcleugh, thunderstorms in Brazil, 130, 147, 191
Calm centre in cyclones, 373
Campbell's sunshine recorder, 58
Capacity correction of barometers, 73, 84
Cape of Good Hope, observed temperature of soil at, 47; winds near the, 284
Capillarity correction of barometers, 84
Carbonic acid, amount of, in atmosphere, 15
Cator's anemometer, 151
Celsius (Centigrade) thermometer scale, 19
'Challenger,' H.M.S., cruise of the, 313
Cherraponjee, rain of, 325

Chronoisothermal diagram for Greenwich, 48; explained, 390
Circulation of atmosphere explained, 240
Cirrocumulus cloud, 126
Cirrostratus cloud, 125
Cirrus cloud, 124; its motion, Ley, 125; Hildebrandsson, 125
Cistern of barometer, 72
Clavering, on St. Elmo's fire, 178
Climate, on change of, with height above sea, 338, 345, 346; insular and continental, 344; chief conditions which affect, 344, 346
Cloud, composition of, 123; levels at which they float, 123; classification of, 124; motion of, 129; amount of, 129; annual and diurnal period of, 129; colours of, 205
Clouston, Dr., on distribution of winds in the Orkneys, 277
Coffin, winds of globe, 163; on relation of wind to pressure, 255
Cold, greatest obtained, 235; compared with heat, 235
Cold currents under warm water, 303; on west coast of Africa, 304
Cold wall of Gulf stream, the, 300
Coldest period of the day, 47; of the year, 48
Colding, on course of Gulf stream, 299
Cole, on evaporation in France, 102
Colouring of clouds, 205
Como Lake, diurnal period of cloud on, 130
Comozants, 178
Compass bearings, correction of, 148
Components of wind velocity, 165
Conductors (lightning), invention of, 167, 183; principle of, 183; erection of, 185; rules for, 186; material of, 185; Dr. Mann on, 186; Conference on, 186
Continental climate, 344
Continuity, equation of, applied to currents, 297
Conversion tables for thermometer scales, 20, 389

398 INDEX.

COR

Corona of the aurora explained, 192
Coronæ explained, 202
Corposants, 178
Corrections for deducing mean temperature from various observations, 38
— of barometers, for capacity, 73, 84; for capillarity, 84; for index error, 84; for temperature, 85, 391; for altitude, 85, 391
— of wind observations from ships' logs, 284
Coxwell and Glaisher, balloon ascent, 12, 66
Croll on ocean currents, 221; on the Gulf stream, 301
Cumberland Lakes, rainfall of the, 330
Cumulostratus cloud, 127
Cumulus cloud, 127
Currents, atmospheric, ascending and descending, their action on precipitation, 138; upper, their existence proved, 243; equatorial and polar, 276
— ocean, Rennell on, 295; drift, 295; stream, 295; their influence on sea temperature, 318
Current, the equatorial, 296, 298, 305, 307; the American arctic, 298; the Gulf stream, 221, 297, 299, 301, 302; the Kuro Siwo, 221, 296, 306; Rennell's, 302; Humboldt's, 299, 307, 344; of Southern Ocean, 298, 309; of South Pacific, 307; the Agulhas, 308; in Polar Regions, 309
Cyclones defined, 365; indraft of air in, 366, 369, 374; localities of, 367; seasonal distribution of, 368; fall of barometer in, 368; waves caused by, 375; motion of, 370, 372, 379; position of greatest danger in, 370; segments of, 370; handling of ships in, 371; calm centre in, 373
Cyclonic system. *See* Depression.

D'ABBADIE, on thunderstorms in Abyssinia, 194

DIU

Dalton's law of mixed gases not true for atmosphere, 112
Dampier, on rain at Gorgonia, 138; on land and sea breezes, 285
Danckelmann, von, on rainfall of Indian Ocean, 327
Daniell's hygrometer, 103
Dark segment, of the aurora, 192
Davos, solar radiation at, 56
Day, length of, in various latitudes, 8
Deep sea temperatures, 310
De La Rue, comparison of lightning with electric spark, 180; on colour of aurora, 193
Density, maximum, of fresh water, 312; of sea water (supposed), 313
Depressions explained, 354; air motion in, 357, 368; contrasted with anticyclones, 357; passage of, described, 358; phenomena on opposite sides of, 359; motion of, 361, 378, 380; subsidiary or secondary, 378
Deviation of winds owing to earth's rotation, 245
Dew, Wells' theory of, 114; explanation of, 118; excessive in tropical climates, 118; annual amount of, 119
Dewpoint, 103, 115
Dines, hygrometer, 104; on annual amount of dew, 119; on influence of height on rainfall, 134; on rainfall in London, 332; on rainfall at Cobham, 334
Distribution of land and water on the globe, 11; its effect on temperature, 219
— of plants and animals, 336, 342
Diurnal range of pressure, 88; at Greenwich, 89; at Calcutta, 89; at Nertschinsk, 89; in Boothia Felix, 89; in British Isles, 91
— — temperature, 37; at Greenwich, 40; dependent on amount of sunshine, 214
— — of cloud, 129
— — of rainfall, 147
— — of wind velocity, 163

DIU

Diurnal range of thunderstorms, 190
Doldrums, rain of the, 325
Dorpat, discussion of wind velocity at, 166
Dove, calculation of dry air pressure, 90, 113; theory of hail formation, 146; isothermal charts, 222; relative temperature of two hemispheres, 230; isabnormal charts of temperature, 232; cause of Trade wind, 244; change of direction of winds, 254; cause of Monsoons, 265; on West Monsoons of the Line, 267; on equatorial and polar currents, 276; weather rules, 349; law of gyration, 360
Drainage, influence of, on the effects of rainfall, 136
Drift currents, 295
Dry air pressure, 90, 113
Dry and wet bulb thermometers, 107
Dryness of air, its effect on radiation, 57, 62
Dublin, annual range of temperature at, 45; evaporation at, 102

EARTH, the form of, 6; orbit of, 7; rotation of, 8; motion of, in its orbit, 8; effect of motion on solar action, 45, 211; condition of its surface in relation to solar radiation, 46; in relation to terrestrial radiation, 60
Eaton (H. S.), on diurnal range of the barometer, 90
Ecliptic, inclination of the, 8
Electricity, atmospheric, 167; laws of, at Brussels, 171; changes of, during thunderstorms, 172
Electric potential, defined, 170
Electrometer, Thomson's, 169
Electroscopes, 168
Equation of continuity, 297
Equator, diurnal range of temperature at, 216; the thermal, 224
Equatorial current, atmospheric, 276
— — ocean, 299, 305, 307
Erman, Adolf, on relation of wind to pressure, 254

FRA

Europe, rain probability of, 336
Evans, Sir F., on currents of Southern Ocean, 298
Evaporation, 95; effect of, 97; difficulty of observing, 97; in India, 99; from various crops, 99; measurement of, 100; Schmid on, 101; in various latitudes (Haughton), 101; in France, 102; in London, 102; in Dublin, 102; compared with rainfall, 102
Everett, on mirage, 206
Exposure, influence of, on terrestrial radiation, 59
Extraordinary rainbows, 201

FAHRENHEIT'S thermometer, 19
Ferrel, on relation of wind to pressure, 255
Firescreens, glass, action of, 53
Fitzgerald, M., on globular lightning, 176
FitzRoy, Admiral, on weather of globe, 363
Five day means for temperature, 43
Floats, fishing, Norwegian, found in Labrador, 309
Floods, connected with drainage, 136
Fog, composition of, 119; formation of, 120; Aitken's theory, 120; density of, in Newfoundland, 120; in Spitzbergen, 122; sea, 122; entry of, in journals, 129
Fogbows (fog-eaters), 201
Föhn, the explanation of, 289; in Greenland, 289
Forecasting, its difficulty, 361
Forests, influence of, on water supply, 136
Forked lightning, 173
Forties, the roaring, 264
Fortin's barometer, 72
'Fox,' drift of the, 309
Fraction of saturation, 110
France, evaporation in, 102
Franklin, invention of lightning

FRE

conductors, 167, 183; proof of nature of lightning, 167
Freezing of salt water, 311; of fresh lakes, 312
Frequency of winds from various points, 276–281

GALES, straight-line, so-called, explained, 377
Galileo, observation of height of water in a pump, 63
Gas explosion, extraordinary effects of a, 156
Gases and vapours, differences between, 95
Gay-Lussac's pipette, 83
Gilbert, Dr., evaporation from various crops, 99
Glaisher, Jas., balloon ascent, 12, 66; on terrestrial radiation, 61; hygrometric tables, 109; factors, 109; rain gauge, 131; on wind pressure, 158; winds at Greenwich, 277
Glasgow, excessive wind pressure at, 158
Glazed frost, 115; its effects in France (1879), 116
Globular lightning, 175
Glories, 203
Godefroy, on glazed frost, 116
Gradients, barometrical, explained, 353
Graduation of thermometers, 21; of barometers, 67
Grass, its influence on terrestrial radiation, 60
Greaves, C., on evaporation in London, 102
Greenland, Föhn in, 289; climate of, 342
Greenwich, diurnal range of temperature at, 40, 216; chronoisothermal diagram for, 48; diurnal range of pressure at, 89; annual range of temperature at, 218; excessive wind-pressure at, 158; distribution of winds at, 276
Guinea current, the, 303
Gulf-stream, the, 221, 297, 318; its

HEA

course, 299; its influence, Croll, 301; Haughton, 301; on the coasts of Europe, 302; in Polar Regions, 309
Gulf-weed, the, 303
Gyration, the law of, 360

HADLEY, theory of Trade winds, 244
Hail, soft, 143; true, 143; insurance, 146; connected with thunderstorms, 190
Hailstones, 143; Abich on, 143; Delcros on, 144; formation of, 145; Volta's theory, 145; Dove's theory, 146
Hair hygrometer, 105
Hall, Basil, ascent of the Peak of Teneriffe, 112
Hallsberg, whirlwind at, 382
Halos, explanation of, 203; relation of, to weather, 205
Hammerfest harbour, 221
Hann, on the causes of rain, 138, 139; on relative temperature of the two hemispheres, 231; on the effect of height on temperature, 236; on prevalent winds and thermal windroses, 280; on hill and valley winds, 288; on the 'Bora,' 383; on rainless regions on land, 328; on influence of mountains on rainfall, 328; on zone of maximum rainfall, 329; on seasonal rainfall of Europe, 333
Harmattan, the, 293
Haughton, Dr., the sun compared with the earth as a source of heat, 51; on evaporation in various latitudes, 101; on effect of rainfall on climate, 140; on motion of currents, 297; on influence of Gulf-stream, 301; on the Kuro Siwo, 306
Heat, amount of, received by each hemisphere, 10; solar, affected by motions of earth, 45, 211; by conditions of surface, 46; variation of, in the daily and annual period, 47; sources of, 50; great-

HEA

est observed, 47, 235; effect of, on pressure, 239; amounts conveyed to atmosphere by land and water respectively, 242

Heat, excessive, compared with excessive cold, 235

Height, influence of, on rainfall, 134; on temperature, 213, 236

Hemispheres, difference of heat received by each, 10; comparison of temperatures, 228, 231

Herschel, temperature of soil at Cape of Good Hope, 47; actinometer, 52; black-bulb thermometer, 53; on terrestrial radiation, 59; on diurnal range of barometer, 90; on radiation fogs, 121; on change of climate with height, 338; Physical Geography, 347

Hildebrandsson, on cirrus clouds, 125
Hill and valley winds, 288
Hill, on zone of maximum rainfall in Himalayas, 329
Himalayas, zone of maximum rainfall in, 329
Hoarfrost, 115, 117; how to foretell, 117
Hooke's anemometer, 150
Horse latitudes, the, 268
Hours of observation for temperature, 41
Howard's classification of clouds, 124; rain-gauge, 131
Humboldt, on excessive dew in South America, 118; isothermal lines, 222; on density of sea-water, 313; current, 299, 307, 344
Humidity, absolute, 110; relative, 110
Hurricanes. See Cyclones
Hutton, hygrometry, 107
Hygrometry, 103; August, 108; Apjohn's formula, 108; Glaisher's tables, 109
Hygrometers, direct, 103; Daniell's, 103; Dines's, 104; Regnault's, 104
— indirect, 105; hair (Saussure's), 105; dry and wet, 107; Leslie's, 107; Mason's, 107
Hypsometry, 21

IAM

ICE, formation of, by radiation, 61, evaporation from, 96; effect of, in reducing range of temperature, 220
Ice clouds, 123
Index error of barometers, 84
India, formation of ice by radiation in, 61
Indian Ocean, rainfall of the, 327
Indraft in cyclones, 366, 369, 374
Insular climate, 344
Insurance against hail, 146
Isabnormal charts, 232; their relation to those of distribution of pressure, 257
Isobars, 256, 354
Isobaric charts, 238; explained, 256
Isothermal charts, 222
— lines, their courses traced, 224–230
Italy, local names of winds in, 293

JAMAICA, land and sea breeze at, 286
James, Sir H., on relation between pressure and velocity of wind, 154
Japanese seas, the currents of the, 306
Johnston (A. Keith, jun.), Range Maps, 339

KÄMTZ, vertical distribution of vapour in atmosphere, 111
Kerguelen, force of wind at, 253
Kew Observatory, certificates for thermometers, 23; for barometers, 85; atmospheric electricity during thunderstorm at, 172
Kew barometer, 73
Khamsin, the, 293
King's barograph, 78
Königswinter, whirlwind at, 382
Köppen, rainfall of the Atlantic, 327; rain probability for Europe, 336; on climatic character of winds, 360
Kuro Siwo, the, of the Pacific, 221, 297, 306, 318

LABRADOR, Norwegian floats found in, 309
Lambert's formula for wind discussions, 161
Lamont, J. von, atmometer, 100;

D D

LAN

on distribution of vapour in atmosphere, 112
Land and sea breezes, 285; Laughton on, 286; Blanford on, 286
Land and water, relative proportion of, on globe, 11; effect of distribution on temperature, 219, 220
Lanes cut by storms in woods, 155
Latent heat, effect of, on temperature, 97, 117. 220
Latitudes, the 'horse,' 268
Laughton, on the horse latitudes, 268; on the winds of the globe, 269; on land and sea breezes, 286; on the existence of a sargasso sea in the Pacific, 306
Law, Buys Ballot's, 255, 353
— of gyration, 360
— of storms, 365
Lawes, Sir J. B., on evaporation from various crops, 99
Leaves, fallen, action of, in checking evaporation, 99
Leh, extreme solar radiation at, 56
Leipzig, discussion of windroses for, Schreiber, 350
Leslie's hygrometer, 107
Levanter, 293
Ley, on cirrus clouds. 125
Liais, estimate of height of atmosphere, 13
Lightning, comparison of, with electric spark, 173, 180; zigzag, 173; sheet, 175; globular, 175; motion of, 174; effects of, 181; return shock of, 181
— conductors, invention of, 167, 183; principle of, 183; action of, 183; erection of, 185; material of, 185; rules for, 186; Conference on, 186
Lind's anemometer, 150
Liverpool, annual period of wind velocity at, 164
London, annual range of pressure for, 93; evaporation in, 102; rainfall of, 332; rain probability for, 331
Looming, 207
Loomis, distribution of aurora, 194; distribution of rain, 322
Lottin, on aurora at Bossekop, 192

MOH

Lunar rainbows, 202
Lustra, for calculation of means, 44
Lyell, Sir C., proportion of land and water on globe, 11, 220
Lying-to in cyclones, 371

MACKEREL SKY, the, 126
Magnetic storms, 196
Magnetism, terrestrial, connection of, with aurora, 196; with sunspots, 392
Mahabuleshwur, rainfall of, 321
Malden Island, rainfall at, 325
Mann, Dr., on lightning conductors, 186
Marco Polo, on monsoons, 248
Mare's tail clouds, 124
Marine barometer, 74
Mason's hygrometer, 107
Mauritius hurricanes, 367, 369
Maury, on rainfall in Trade wind zone, 327
Maximum density of vapour, 96
— thermometers, Six's, 25; Phillips', 27; Negretti's, 27; hours for observing, 42
Mean sea level, 85
— temperature, determination of diurnal, 38; five day, 43; monthly, 43; annual, 44
Meldrum, on Mauritius cyclones, 369; on indraft of wind in cyclones, 369, 375; on connection between sunspots and weather, 393
Meteorology contrasted with astronomy, 1; different lines of study of, 3; cosmical, 4
Minimum thermometers, Six's, 25; Rutherford's, 29; Casella's, 29; grass, 61; hours for observing, 42
Mirage, 206
Mist, composition of, 119, 122
Mistral, the, 292. 383
Mock suns, &c , 203
Mohn, Prof., on distribution of vapour tension, 111; on classification of thunderstorms, 189; isothermal charts, 222; on causes of barometrical oscillations, 248; on sea temperature, 314

MON

Monsoons, explained, 247, 265;
  Dove on, 265; Blanford on, 265;
  Aristotle on, 266; the Arab
  writers on, 266
— the West of the Line, 266
— the soft place in the, 268
Mont Blanc, observations of solar
  radiation on, 57
Moon, heat from the, 51; mock, 203
Motion of clouds, 129
Mountains, effect of, on temperature, 236, 338; on rainfall, 328
Müller, Hugo, comparison of lightning with the electric spark, 180;
  on colour of aurora, 193

NAGPOOR, evaporation at, 99
  Negretti's maximum thermometer, 27
Nertschinsk, diurnal range of pressure at, 89
Newfoundland, banks of, fogs on, 120
New Haven (Conn.), annual range of temperature at, 45
New South Wales, temperature of soil observed in, 47
New Zealand, winds of, 289
Nimbus cloud, 128
Noah's ark in the sky, 125, 359
Noise of the aurora, 192
Nordenskiold, on snowdrift in Siberia, 326
Nortes, the, 294, 383
Northumberland Sound, summer temperature of, 236
North-west wind, in cyclonic systems, 190, 359, 376
North-westers, of New Zealand, 289
Norwegian fishing floats found in Labrador, 309

OETTINGEN, VON, on wind discussion, 165
Optical phenomena, 199
Orbit of the earth, 7
Orkneys, the, annual period of wind velocity in, 164
Osler's anemometer, 150

PRE

Oxygen in atmosphere, variations in amount of, 15
Ozone, its discovery, 196; its nature, 197; difficulty of observing, 197; Dr. Tripe on, 198

PACIFIC OCEAN, limits of trade winds in, 246; rainless regions in, 325
Pallium, a form of cloud, 126
Pamperos, 294, 383
Papagayo winds, 383
Parhelia and paraselenæ, 203
Parmenides, division of earth into zones, 5, 338
Parry, Sir E., currents of Polar Regions, 309
Pascal, proof of the principle of the barometer, 65
Peruvian current, the, 299
Phillips' maximum thermometer, 27
Physical geography, works on, 347
Piddington, Law of storms, 365; on indraft of air in cyclones, 366
Pike's Peak, summer temperature at, 237; winter and summer pressures at, 241
Pipette, Gay-Lussac's, 83
Pitlekaj, snowdrift at, 326
Planets, the classification of, 6; laws of motion of, 7
Plants, distribution of, 338, 342
Pocky cloud, the, 128
Poëy, M., on clouds, 126
Polar currents, 276
Poles of cold (Brewster), 234
'Porcupine,' H.M.S., cruise of the, 313
Potential, in electricity, defined, 170
Precipitation, its various forms, 114, 119
Pressure, atmospheric, 63; its average amount, 64; diurnal range of, 88; Dove, 90; Eaton, 90; Buchan, 91; Simmonds, 92; Strachan, 92; influence of vapour on, 92; annual range of, 93; affected by heat, 239; reasons for increase or decrease, 249; causes of excessive, 249; regions

PRE

of defective, 250, 257; explanation of defective, at South Pole, 251; relation of, to wind, 253, 353; distribution of, 256; relation of, to isabnormals of temperature, 257
Pressure, wind, measurement of, Hooke, 150; Wild, 150; Lind, 150; Osler, 150; Cator, 151; excessive at Bidston, 156; at Calcutta, 156
—— and velocity, relation between, 154
Primary rainbow, 200
Probability of rain, 334; in British Isles, 334; in Europe, 336
Psychrometer (dry and wet bulb thermometer), the, 107
Pump, action of the, 64
Pumping of barometers, 75
Punos winds. 291
Purga, the, 294

QUETELET, discussion of atmospheric electricity at Brussels, 171

RADIATION, solar, 51; observation of, 55; affected by vapour, 113, 212; affected by angle of incidence, 211
— terrestrial, 59; observation of, 60; affected by vegetation. 60, 212; Glaisher on, 61; affected by vapour, 113, 212
Radiation fogs, 121
Radii of the Earth, the, 6
Rainbow, 199; explanation of, 200; primary, 200; secondary, 200; supernumerary, 200; extraordinary, 201; in connection with weather, 201; lunar, 202
Rain-probability defined, 334; for Valencia, 334; London, 334; Cobham, 334; different for different amounts. 335; of Europe. 336
Rainfall, compared with evaporation, 102; influence of height above ground on, 133; its amount over large areas. 135; instances of excessive falls, 135, 137; storage

RON

of, 135; influence of cultivation on its effects, 136; average amount in United Kingdom. 137; causes of, 137–139; effects of, in beating air, 140; diurnal period of, 147; at Mahabuleshwur, 321; Batavia, 323; Buitenzorg, 323; Cherraponjee, 325; Ascension, 321; Malden Island. 325; Santiago, 326; Valdivia. 326; distribution of, 321; influence of mountains on, 328; seasonal distribution of, 330; in British Isles. 332; in London, 332; in Europe, 333
Rain-gauges, Howard's, 131; Glaisher's, 131; The 'Snowdon,' 132; Beckley's, 132; diameter of, 132; exposure of, 133
Rainless regions, on land, 326, 328; at sea, 328
Range of pressure, diurnal. 89; Herschel on, 90; annual, 93
— temperature, diurnal, 37, 214; annual, 45; maps of, Keith Johnston, 339; Supan, 340; laws of, 341
Rarefaction over Central Asia in summer, 93, 259
Réaumur's thermometer scale, 20
Redfield, law of storms, 365; on indraft in cyclones, 366
Redier's barograph, 77
Reduction of barometers to 32°, 85
—— to sea-level, 85; theoretical difficulty of, 87; simple rule for, 86
Regnault's hygrometer, 104
Reid, law of storms, 365
Relative humidity, 110
Rennel, on currents, 295; his current, 302
Renou, on history of thermometer, 17
Rensselaer harbour, temperature of, 236
'Resolute,' H.M.S., drift of the, 309
Resolution of wind velocities, the, 165
Resultant winds, deceptive, 163–165
Richmann. Prof., death of, 168
Roaring Forties, the, 264
Robinson, Dr., on influence of height on rainfall, 135; anemometer, 151
Rolling of thunder explained, 178
Ronald's barograph, 77

Ross, Sir J. C., on defective pressure near South Pole, 252; force of wind at Kerguelen, 253; on deep sea temperatures, 313
Rosser, diagrams of storms, 371
Rotation of the earth, its influence on temperature, 8, 211; on motion of air, 245
Rundell, on winds at Liverpool, 162
Russian Empire, Wild's temperature charts for, 222
Rutherford's thermometer, 29; defects of, 30

SABINE, on St. Elmo's fire, 178; on aurora at low levels, 193; on the Guinea current, 303; on relation of sunspots to magnetism, 392
St. Elmo's fire, 178
St. Théodule, Col of, annual range of temperature at, 236
Sandwick Manse (Orkneys), annual period of wind velocity at, 164
Santiago (Chili), rainfall of, 326
Sargasso Sea, the, 302; in South Atlantic, 305; in Pacific, 306
Saturation, fraction of, 110
Saussure, hygrometer, 105
Scale, Beaufort's, for wind force, 158
Scales of barometers, best material for, 66; of thermometers, different, 19
Schmid, on evaporation, 101
Schott, on velocities corresponding to degrees of Beaufort's wind scale, 160
Schreiber, wind rose for Leipzig, 350
Schück, Capt., on indraft of wind in cyclones, 374; on arched squalls, 384
Schwabe, sunspot measures, 392
Scirocco, the, 292
Scoresby, on solar radiation in the Arctic Seas, 57
Scottish Meteorological Society, on sea temperatures, 319
Sea breeze, the, 285
Sea fogs, 122
Sea-level, on reduction of barometer readings to, 85
Sea surface temperatures, 314; regulated by currents, 318; on coasts of the British Isles, 319
Seasonal distribution of rainfall, 330
Secondary rainbow, 200
— depressions, 380
Setting the barometer, 68
Sheet lightning, 175
Ships' logs, on treatment of wind observations from, 282
Siberia, snow drift in, 326
Silver thaw, 115
Simoon, the, 293, 382
Siphon barometers, 75
Six's thermometer, 25
Sky, appearance of, during thunderstorms, 188
Sleet, 143
Sling thermometers, 36
Smith, Dr. Angus, on impurities in atmosphere, 14; on importance of temperature to health, 17
Smoke, use of, to ward off hoar frost, 117
Snow, evaporation from, 96; indirect influence of, on barometrical pressure, 249; measurement of, 141; amount of, 142; crystals of, 142
Snowdrift, in Arctic regions, 326
Soft hail, 143
Soft place in the monsoon, the, 268
Soil, excessive temperature of the, 47
Solano, the, 293
Solar radiation, 51; excessive, at Leh, 56; at Davos, 56; Violle's results, 57
— system, the, 6
Somerville, Mrs., 'Physical Geography,' 347
South Pole, reduced atmospheric pressure near, 252
Southerly 'busters' of New Zealand, 290
Spirit thermometers, 29; defects of, 30; construction of, to ensure sensitiveness, 31
Spitzbergen, fogs in, 122
Sprung, on rainfall of Atlantic, 327
Squalls, 384; arched, 384
Standard pressure for thermometer graduation, 22

STA

Standard barometers, 72
Stations of first, second, and third orders, definition of, 41
Stevenson's thermometer screen, 34
Stewart (Balfour), actinometer, 52; on influence of sunspots on weather, 393
Stokes, Prof., sunshine recorder, 59; on the theory of Robinson's anemometer, 152
Storms, law of, 365; revolving (see Cyclones)
— — of temperate zone, their nature, 376; their motion, 378
Storm wave, the, 375
— warnings, 379
Strachan, R., rule for reduction of barometer to sea-level, 86; diurnal range of pressure, 92
Strachey, Gen., distribution of vapour in vertical height, 112, 329
Stratus cloud, 126
Stream currents, 295
Sturt, Capt., temperature of soil in New South Wales, 47
Subsidiary depressions, 380
Subtropical rain region, 331
Summer, relative length of the, in the two hemispheres, 10, 231; lightning. 175; temperature, the lowest observed, 236; rains, regions of, 331
Sun, the heat from, affected by motion of earth, 45, 211; by angle of incidence, 211; influence of, on diurnal range of temperature, 214
Suns, mock, 203
Sunshine recorder, 58
Sunspots, relation of, to weather, 363, 392; periodicity table of, 394
Supan, Prof., proposed new classification of zones, 339; range maps, 340; range laws, 341
Supernumerary rainbows, 200
Symons, Rainfall system, 135; on influence of height on rainfall, 134
Sympiezometer, 81

T AIT, Prof., on the rolling of thunder, 179; on the cause of

THE

intensity of electrical phenomena in thunderstorms, 180
' Tegethof,' drift of the, 309
Tehuantepec winds, 383
Teisserenc de Bort, on relation of isabnormals of temperature to pressure, 257
Telegraphy of weather, 362
Temperature, diurnal range of, 37; affected by state of sky, 214; mean diurnal, determination of, 38; from maximum and minimum readings, 42; mean annual, 43; of the soil, excessive, 47; influenced by height, 213; annual range of, 216; at high levels, 238
— deep sea, 311; sea surface, distribution of, 314
— charts, isothermal 223
— isabnormal, 232; relation of, to barometrical pressure, 257
Teneriffe, ascent of Peak by Basil Hall, 112; annual change of wind on, 243
Tension of vapour, 110
Terrestrial magnetism, connected with aurora, 196; relation of, to sunspots, 393
— radiation, 59
Thaw, silver, 115; the 'up-bank,' 213, 345
Thermal equator, the, 224
Thermographs, mechanical, 22; photographic, 33; records of, 37;
Thermometer, defined, 17; history of, 18; best liquid for, 19; different scales, 19; graduation of, 21; verification of, 23; ordinary, range of, 23; displacement of zero in. 24; requisite accuracy of, 23; self-registering, 25; Six's, 25; maximum, 26; setting of, 28; minimum, 29; separation of spirit in, 30; exposure of, 33; screens, 34; sling (thermomè re fronde), 36; black bulb *in vacuo*, 54
Thermometer attached to a barometer, 70
— deep sea, 310
— conversion tables for, 389

THO

Thomson's, Sir W., electrometer, 169
Thunder, its nature, 178; distance it is heard, 179
Thunderstorms, their effect on atmospheric electricity, 172; their nature, 172, 189; appearance of sky during, 188; 'heat,' 189; 'cyclonic,' 189; diurnal and annual period of, 190; in Arctic regions, 191; in Abyssinia, 191
Tornadoes, defined, 381; described, 384
Torricelli, invention of the barometer, 63; his vacuum, 65
Toynbee, Capt., on cold water near Equator, 304; on climatic character of winds, 360; on indraft of wind in hurricanes, 374
Tracks of hurricanes, 370, 372
Trade winds explained, 244; Hadley's theory, 244; limits of, 246, 262; described, 261; direction of, 262; South-east, interruption in, 263; zone, not rainless, 327
— Laughton on, 269
Tramontana, the, 292
Trevandrum, range of temperature at, 215
Tripe, Dr., on globular lightning, 176; on ozone observations, 198
Tyndall, Prof., on action of vapour in checking radiation, 113
Typhoons, 365

UNITED STATES, the, weather telegraphy in, 362
Upbank thaw, the, 213, 345
Upper currents of atmosphere near equator, 243

VACUUM, Torricelli's, 65
Valdivia, rainfall at, 326
Vapour, maximum density of, 96; action of, on radiant heat, 113
— Geographical distribution of, 111
— vertical distribution of, Kämtz, 111; Strachey, 112, 329; Basil Hall, 112; v. Lamont, 112
— Tension, 110

WEA

Vapours and gases, difference between, 95
Varenius, on Trade winds, 244
Variable winds, the, 267
Variation of the compass, 148
Veering of wind, 361
Vegetation, influence of, on radiation, 61
Velocity anemometer, Robinson's, 151
— and pressure of wind, relation between, 154
Verification, of thermometers, 23; of barometers, 85
Vernier, for barometers, explained, 67
Violle, M., on absorption of heat by atmosphere, 57
Volcanoes, ashes from, borne by upper currents, 244
Volta, theory of hail, 145; mode of collecting atmospheric electricity, 169

WALL screen, for thermometers, 35
Walmer whirlwind, the, 381
Warmest period of the day, 47; of the year, 48
Warnings of storms, 379
Water, proportion of, on earth's surface, 11, 219; effects of, in reducing range of temperature, 220, 342; in heating atmosphere, 242
Water, fresh, maximum density of, 312
— salt, freezing of, 311, 313
Waterspout produced by a whirlwind, 382
Wave, the storm, 375
Weather, connection of aurora with, 195; rules for, Dove, 349; Schreiber, 350; dependent on cyclones and anti-cyclones, 352; in cyclones and anti-cyclones respectively, 358; relation of, to sunspots, 363, 392
— charts, 355, 356
— forecasting, 361
— galls, 200
— telegraphy, 36

WEI

Weighting wind observations from ships' logs, 283
Wells, theory of dew, 114
Werchojansk, winter temperature of, 228, 235
West Monsoons of the Line, the, 267
Wet bulb thermometer, management of, 107
Weyprecht, on connection between aurora and weather, 195
Wheatstone, measurement of velocity of lightning, 174
Wheel barometer, the, 76
Whirlwinds, 381; Walmer, 381; Hallsberg, 382; produce waterspouts, 382
White squalls, 384
Whitley, N., on sea temperature on British coasts, 319
Wijkander, on fogs in Spitzbergen, 122; on connection between aurora and magnetism, 196
Wild, anemometer, 150; temperature charts of Russian empire, 222; relation of isabnormals of temperature to pressure, 257
Wind, 148; direction of, 148; pressure and velocity, relation between, 155, 158
— pressure, excessive, at Bidston, 156; at Calcutta, 156; at Greenwich, 158; at Glasgow, 158
— pressures, table of, 157; force, Beaufort's scale, 158
Wind, discussion, 161; Coffin, 163; Wojeikof, 163
— velocity, diurnal range of, 163; annual range of, in British Isles, 164
— action of change of level on, 288
— relation of, to barometrical pressure, 253, 353
— north-west, in a depression, 190, 359
— veering and backing of, 361
— climatic character of, depends on origin, 360
Wind and current charts (Admiralty), 282, 314

ZON

Wind galls, 200
Wind gauges, pres ure, Hooke, 150; Wild, 150; Lind, 150; Osler, 150; Cator, 151
— velocity, Robinson, 151
Winds, the Trade, explained, 244, 262; the Anti-trades, 264; permanent, 261; periodical, 264; variable, 267; Laughton on, 269; distribution of, in British Isles, 276; general principles of their motion in the temperate zone, 275; climatic influence of, 278, 280, 350; local names for, in Italy, 295; at sea, difficulty of discussing, 282; land and sea breezes, 285; hill and valley, 288; the Föhn, 289, 291; the Punos winds, 291; Scirocco, 292; Tramontana, 292; Bora, 292, 383; Mistral, 292; Levanter, 293; Harmattan, 293; Khamsin, 293; Solano, 293; Simoon, 293, 382; Buran, 293; Purga, 294; Pamperos, 294, 383; Nortes, 294, 383; Papagayo and Tehuantepec, 383
Wind-roses, 166; thermal for northern hemisphere, 278, 280; weather, 349; for Leipzig, 350
Winter rains, regions of, 331
Wise, T. A., on formation of ice by radiation in India, 61
Wojeikof, discussion of winds of globe, 163; on causes producing excessive pressure, 249; on rain distribution, 322, 324; on rainfall of Java, 322
Wolf, Rudolf, table of sunspot-periodicity, 394
Woods, influence of, on rainfall, 136

ZERO, displacement of, in thermometers, 24
Zigzag lightning, 175
Zones, geographical definition of, 10; proposed new classification of, 339

www.ingramcontent.com/pod-product-compliance
Lightning Source LLC
Chambersburg PA
CBHW032140010526
44111CB00035B/630